'Wine should be a pure reflection of its soils and of the entirety of its environment. We should question everything that takes us further away from that fact.'

NICOLAS JOLY
VINEYARD OF THE COULÉE DE SERRANT

Above: The vines at Lightfoot & Woolfville in Canada.

Opposite: The mist rolls across Spring Mountain
at Cain Vineyard & Winery in Napa.

Wine Revolution

The World's Best Organic, Biodynamic & Natural Wines

Jane Anson

First published in 2017 by
Jacqui Small
An imprint of The Quarto Group
74–77 White Lion Street
London N1 9PF

Publisher: Jacqui Small
Managing Editor: Emma Heyworth-Dunn
Senior Commissioning Editor: Fritha Saunders
Assistant Editor: Joe Hallsworth
Designer: Luke Fenech
Editor: Hilary Lumsden
Photography: Kim Lightbody, pages: 10, 13, 14, 17, 21, 27, 31, 42–3, 80–81, 122-123, 174–5, 224–5.
Production: Maeve Healy
ISBN: 978-1-911127-29-1

A catalogue record for this book is available from the British Library.

2019 2018 2017
10 9 8 7 6 5 4 3 2 1

Printed in China

Above: Elisabetta Foradori getting hands-on in her Italian winery.

Farm to Table Wines

Chez Panisse opened its doors in downtown Berkeley on August 28, 1971. The set menu for the first night, priced at US$3.79, was *pâté en croûte*, salad, duck with olives and almond tart. The first plates of food didn't make their way out of the kitchen until nearly 9pm, and it would take a few weeks before things really settled in. But in this two-storey Victorian townhouse, founders Alice Waters and Lindsey Shere ushered in the foodie, farm-to-table movement that has since been championed by Dan Barber, Gaston Acurio, Hugh Fearnley-Whittingstall, Umberto Bombana... all rock-star chefs who produce delicious food in touch with nature, with the seasons, with the land.

The idea of buying locally, cooking with seasonal ingredients, supporting responsible farming has become so accepted as to barely raise an eyebrow. And yet when it comes to wine, it is still considered geeky and kind of pointless to care about the same thing. After all, aren't all grapes organically grown in a field somewhere?

Well, the short answer to that is no. Just like much of the food we eat, plenty of wine is produced for a mass audience, with shortcuts taken along the way to ensure that they taste good without costing a fortune to make.

So, shouldn't we start celebrating the winemakers who buck this trend, and instead apply the Chez Panisse philosophy to their vineyards? The ones who treat their workers fairly, reduce their carbon emissions, farm without pesticides? Or those who plant hedgerows to encourage biodiversity, use grapes that are indigenous to their regions and add as little as possible during the winemaking process?

There are plenty of them out there. Aubert de Villaine, Elisabetta Foradori, Pepe Raventós, Jean-Laurent Vacheron, Olivier Humbrecht, Eloi Durrbach, Christine Vernay, Nicolas Joly, David Paxton... these are winemakers who should be talked about in the same breath as chefs like Waters, Barber and the rest.

That's what this book is about – a celebration of these committed, dedicated producers. If their wines are in here, it is because they taste

Opposite: As with all artisan winemakers, Ray Nadeson at Lethbridge in Victoria is hands on in both vineyard and cellar.

OLL CH

brilliant, will enhance what you're eating and provide a moment of shared happiness with whoever you are drinking them with. But they also all come with a story, from people who care about authenticity, and want to preserve the land that nourishes their grapes.

BIG VERSUS SMALL

I've been living in France since 2003. Shopping seasonally, going to local markets, buying direct from farms has slowly but surely become second nature. It has brought me closer to the food I eat, and it's fun. Extending that way of thinking about food to wine seems like the best way to capture that enjoyment in our glass. And it's worth pointing out that I live in one of the biggest wine regions of them all; Bordeaux.

This is a place where the image is painted of men in suits making polished wines. You'll find several Bordeaux wines in this book that are the opposite of that idea, and there are plenty more that I didn't have the space to include. But, living here has helped me to understand that authentic wine is not as simple as big versus small. It's why you'll find some of the world's biggest and most prestigious names here, alongside some of the smallest single-person operations. What's important is that each winemaker supports an idea of farming that is respectful of the future, and that looks to capture a snapshot of time, place and culture in a glass of wine.

It's not a definitive list – how could it be? I have had invaluable help and advice from some of the greatest, most switched on sommeliers around the world, who understand exactly why we should think about artisan wines in the same way as we do about farm-to-table food. They have helped me, along with many wonderful consultants, retailers and friends, to discover many of these amazing wines that I am so excited to share with you. Throughout the book you will find advice from them on food styles that bring out the best in the chosen wines.

Deciding who to include in the final list has been thrilling, challenging and enormously tough, and I apologize right now for any missing names. But that's the magic of wine; there is always something new to discover.

Opposite: Nebbiolo at G.D. Vajra in Piedmont.
Above: Pepe Raventós ploughing with horses at Raventós i Blanc in Spain.

Food & Wine

Besides the artisan, handcrafted method of production, there's another reason these are farm-to-table wines. It's because that's where they belong – on a table over lunch or supper, or served outside with a picnic, or in a bar with a few local tapas or sharing-plate specialities.

Food and wine are natural friends, and finding the right combination can heighten the pleasure of both. Pairing them can be as easy or complex as you like, and you'll find suggestions throughout this book from sommeliers who work in some of the world's leading restaurants and bars. In some cases I have also asked the producers for their recommendations, or simply suggested pairings that I have found work particularly well.

Balancing intensity and weight of flavours is one key thing to bear in mind, as is finding the right balance between, for example, sweetness in wine and spice in food. Certain tricks can heighten the enjoyment of both elements, such as using acidity in wine to cut through richness in food.

Each chapter opens with a sommelier sharing their advice and experience. But one element that is crucial but often overlooked in food matching is alcohol level, and it is something I have thought about when deciding where the wines should sit within the book. Alcohol plays an important role in the body, structure and weight of a wine, all of which come into play when pairing with food.

Sparkling & Fresh, Crisp Whites tend to be lighter in alcohol, and the wines in Section One range from 11% ABV to 13.5% ABV, while the Rich & Round Whites in Section Two can head up to 14.5% ABV. Looking at the reds, Section Three's Light & Sculpted wines start at 11% ABV, rising to just above 13.5% ABV, while the Full & Warming Reds of Section Four go up to 15.5% ABV in a few instances, with an average of 13.5% to 14.5% ABV.

It's not an exact guide to how they will taste, because balanced wines have the alcohol fully in tune with the other elements (and I've only picked ones that have got that trick exactly right) but it's worth thinking about when food matching, or simply when picking wines for different occasions. The perfect supper bottle on a winter's evening, for example, might be all wrong for a summertime lunch when you have to head back to work afterwards.

The structure of the book makes the alcohol consideration easy to work out.

Opposite: Think about matching style, weight and flavours of food and wine.

How The Book Works

The winemakers here all work either organically, biodynamically or in ways that specifically and consistently help to minimize the impact on the environment in which they grow and make their wines. I have tried where possible to include only producers with a track record. There are many small winemakers doing weird and wonderful things the world over but sometimes the results are a bit hit and miss. Most are certified by an official body for their green credentials, but some are not, either because they don't feel the need to do so or because their way of working – as with many natural wines – is not yet officially recognized. But, I have tried (except in a few clear instances, which are marked with the Low Intervention symbol, see right) to ensure that they are applying these practices across their entire estates, not just picking a few hectares (ha) for a marketing push.

NAVIGATING THE WINES

My main selection criteria is that the producers make wines that are terroir-led, that they are engaged in true sustainability – meaning they ensure microbial health of the soils and the vineyard's ability to thrive for the next generation of winemakers – and that the resulting wines have vibrancy and personality. Some of the producers have extended the philosophy to ensuring their wineries are 'off-the-grid' and entirely self-sufficient in energy production (see p.34). There are tasting notes for all the wines, and profiles on producers who have been particularly influential in their regions, or are doing something especially unusual in their winemaking.

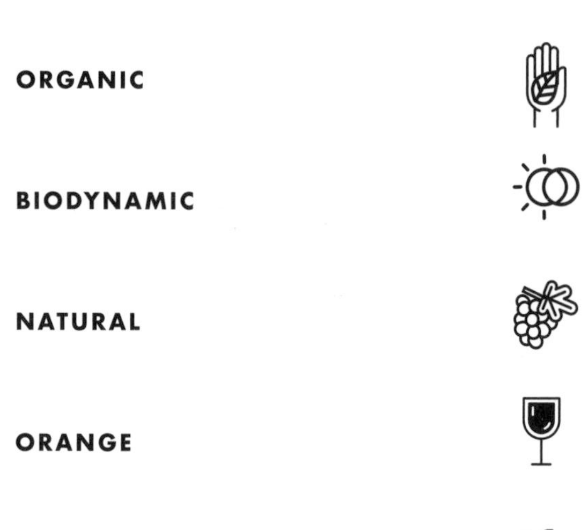

The list of allowable practices for organics and biodynamics is extensive, and changes slightly from Europe to the US to Asia. For further reading on this, I have listed the official bodies at the back of the book on p.256. In the following pages I summarize the basics behind each category to give you the top line, and each wine note in the book includes a symbol indicating the category or certification the wine falls under (see above). And finally, because wine is best when it is part of a meal, I have split the wines into tasting categories by style, and asked a sommelier to give some food-matching tips at the start of each section. These guys are the experts, and I love listening to them talk about how to pick the right ingredients to set these wines alive. I hope you enjoy the results – and are inspired to try plenty of your own.

Opposite: Artisan, handmade wines come in all styles and flavours.

Vins certifiés Agriculture Biologique et contrôlés Bio-Dynamique
BIODYVIN
AB
FR-BIO-01
Agriculture France
Opok 2015
"Opok", dieser traditionelle, regionaltypische Begriff, bezeichnet die Böden hierzulande. Und Opok ist namensgebend für diesen bodenständigen Wein. Der kalkhältige Mergel prägt ihn, macht ihn einzigartig und unverwechselbar. Im Opok sind die Sorten Welschriesling, Sauvignon, Morillon und Muskateller vereint. Lebendig. Nachhaltig. Empfohlene Trinktemperatur 14 C.
Weißwein aus Österreich
Abfüller: Weingut Maria und Sepp Muster, A-8463 Leutschach
enthält Sulfite (nur natürlich gebildete)
contains sulfites (only natural occuring)
Product of Austria
www.weingutmuster.com
Etikett: Horizonte Beppo Pliem
trocken, L 04/17
AT-BIO-301
AT-Landwirtschaft
demeter
Biodynamischer Wein
750 ml
11 %vol
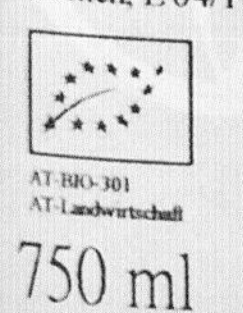

Wine Labels

Different countries, different regions, different producers will all have their own approach to labels. Some are hugely traditional, some innovative, some handed down from older generations, some just a personal reflection of the current winemaker's personality. Yet all are handcrafted, authentic wines that perfectly fit the philosophy of drawing back the curtain between vineyard and wine glass, and respecting the environment at every step of the process.

As a rule, it's the back label where you will find the most interesting information. Not all countries have the same regulations about exactly what information must be included. To date the regulations are less strict than for food, but there are plenty of producers lobbying for the change, and perhaps the natural wine movement will have an impact here, as they champion the fact that they are additive-free.

If a wine is certified organic or biodynamic, that information will appear via a logo from an official certification body (see p.24 for the list of key certification groups). Other labels will point out if the wine is vegetarian or vegan-approved, again with the logo of the relevant certification body. Increasingly wine labels have a QR code, which will give further information when scanned by a phone – including serving suggestions, ideal temperature, even music playlists. Particularly prestigious wines have authentication codes to guard against fraud and counterfeits – you simply enter the code on the bottle into the producer's website and receive a confirmation that the bottle you have is the real thing.

Opposite: The back labels of wine bottles contain mandatory and optional information.

Organic Wine

Defining what organics means in winemaking is deceptively easy. Just as with organic farming of fruits and vegetables, its proponents look after their vineyard without the use of man-made chemicals. In other words, organic wines are made from grapes grown without artificial fertilizers, weedkillers, fungicides or pesticides. That doesn't mean they aren't allowed any help against the weather – organically farmed grapes can use copper, sulphur dioxide (SO_2) and vine treatments made from vegetable and mineral extracts.

In the winery, producers must stay away from many of the 50-or-so additives that are available in more conventional systems of production, unless they are derived from raw materials of agricultural origin that should be in organic form wherever available. The limited selection of additives allowed must also not modify the essential nature of the wine.

Opposite: Organic wines are made from grapes grown without chemical fertilizers or weedkillers.

OCCHIPINTI
SP68
D'AVOLA E FRAPPATO
TERRE SICILIANE
INDICAZIONE GEOGRAFICA TIPICA
2 0 1 5
Imbottigliato all'origine da Az. Agr. Arianna Occhipinti
Vittoria - Italia - RG/3601/IT - Prodotto in Italia
750 ml e
13 % vol.
ANGIOLINO MAULE
MASIERI

The final wines must contain lower levels of sulphites – a common food preservative that prevents wine from spoiling after fermentation – than conventional wines (although exact levels differ depending on the region of production, see p.37, as do the allowable list of products in vineyard and cellar). However, organic regulations in the EU do allow certain processes such as heat treatments, filtration and reverse osmosis because 'at present no alternative techniques are available to replace them', but it is made clear 'their use should be restricted'.

ORGANIC AROUND THE WORLD

The modern push towards organic viticulture dates back to the 1950s, going hand in hand with the rise of big-business farming, which promoted high yields helped along by synthetic fertilizers and chemical or synthetic disease protection. Although the first applications of nitrogen to plant roots in the form of ammonia date back to the 1840s, the real movement towards quantity over quality had its roots in the widespread food and labour shortages after World War II, but this essential push towards mechanization and mass agriculture soon became part of a global agro-chemical industry, which continues to make vast profits from farmers' reliance on these products.

There were always pockets of resistance to this, which slowly but surely became more organized. The first national organic body, the Association Française pour l'Agriculture Biologique, was created in 1962, and Julien Guillot of Domaine des Vignes du Maynes is the son of the winemaker who was behind the lobbying for this (see p.143). European-wide regulations for organic grape production were first introduced in 1991, and the rules were extended to the cellar and wine-making practices in 2012. Since then, certified organic wines made in the EU use a green leaf sign on the label, indicating a wine recognized as an organic product. The rules operate at a

Above: Clos des Vignes du Maynes is one of the organic pioneers in France.

European level but many of the member states have their own certification body or bodies at a national level – which is why you might see the Certifié Agriculture Biologique label in France or Bio-Siegel in Germany, among others (France has six government-approved certification bodies). Over in Australia, the main certification body, the National Association for Sustainable Agriculture, uses NASAA Certified Organic. In the US, the Department of Agriculture's National Organic Program uses USDA Organic. More details of certifying bodies can be found on p24.

The choice isn't always easy of course. Losing a crop to disease or bad weather is devastating financially as well as emotionally. As Deirdre Heekin of La Garagista winery (see p.46) in Vermont writes in her book *An Unlikely Vineyard*, 'I understand daily that the word sustainable must embrace economics as well as the environment'. But she points out that the constant application of chemicals provides only short-term solutions and results in vineyards that can not sustain themselves. She says, 'We will not be enticed away from our belief that working with nature is essential and fighting against nature is foolhardy.' Heekin is one of many US winemakers included in this book, and you will find organic producers from all over the world celebrated here. Indeed one of the fathers of organic farming, English botanist Sir Albert Howard, began work as an agricultural advisor in India in 1905, before advising the Soil Association in the UK and what became the Rodale Institute in the US.

It should not be surprising to see so many European wines among this list. Between 2002 and 2011, according to Eurostat figures, Spain's area of vineyards under organic cultivation jumped from 16,000 ha to almost 80,000 ha, while France's share went from 15,000 ha to 61,000 ha and Italy's from 37,000 ha to 53,000 ha. Between them, these three countries account for 73% of world organic production.

Above: Julien Guillot at Clos des Vignes du Maynes works in traditional, sustainable ways.

Biodynamic Wine

Where organic farming is essentially a continuation of traditional agricultural methods but without the chemicals, biodynamics (from the Greek *bios* for life and *dynamikos* for powerful) moves things on to another level. One of its leading proponents (and a man whose name just kept on coming up when I asked producers about their inspiration) Nicolas Joly describes biodynamics like, 'tuning a radio. We are tuning the vine to the frequencies that bring it to life'.

The method was founded on principles developed in the 1920s by the son of an Austrian railway chief, Rudolf Steiner. He was an architect by trade, but seems to have been restlessly interested in any number of subjects. In 1924 he gave a series of lectures to farmers promoting the use of natural preparations to help their plants develop healthy immune systems. His ideas had a lot in common with organic farming – the use of livestock manure to help plant growth, for example, and using cover crops and crop rotation to revitalize soils, along with the encouragement of polyculture and biodiversity.

Opposite: Biodynamic wines are renowned for the vibrancy and clarity of their flavours.

KELLEY FOX WINES
2015
MIRABAI PINOT NOIR
TERRES CHAUDES
CHAMPIGNY
2015
FORADORI
FORADORI
2014

But he went further, in connecting the forces of nature and the planets with the health and vitality of plants (it could be said that Isaac Newton got there first, with his discovery that tides were affected by the gravitational pull of the moon, but Steiner drew on traditional farming techniques dating back to Greeks, Romans and Egyptians). Steiner's lectures were later published as *What is Biodynamics? A Way to Heal and Revitalize the Earth*.

BALANCING ECOSYSTEMS

In biodynamics, agricultural landscapes are viewed as connected ecosystems where the balance of every element is essential to the whole. The aim is to create a self-sufficient environment and to farm with respect to the cycles of the sun and moon, which is why you will find a lunar calendar hanging on the walls of any biodynamic winery that you happen to walk in to.

Home gardeners often use these too, making use of the idea that certain days are better for planting, and others for clearing weeds and other tasks, all depending on observations, such as plants absorbing more water during a full moon.

The nine natural preparations, or infusions, that are applied to vines under the biodynamic (BD) system are derived from plants, such as nettles, dandelions, yarrow, valerian root and chamomile blossoms or from fermented herbs, minerals and cow manures. These are used either as field sprays or for compost preparation, dynamized in water through a special stirring technique, and applied in minute doses as with homeopathic remedies in natural medicine. The field spray BD preparations are listed as 500 (horn manure), 501 (horn silica) and 508 (horsetail plant); the compost BD preparations as 502 (yarrow), 503 (chamomile), 504 (nettles), 505 (oak bark), 506 (dandelion) and 507 (valerian). One of the lovely side effects I have observed is that you often find these plants begin to grow naturally among the vines in biodynamic vineyards.

As with organic farming, biodynamics still allows the use of copper sulphate to reduce the risk of mildew, but producers are often sparing in their use,

Opposite and above: The lunar calendar has been used through the centuries to help farmers and sailors.

as there are risks to the microflora of the soil with copper. And if you've heard of one thing to do with biodynamics, it is probably that cow horns filled with fermented cow manure are buried in the vineyards to encourage soil fertility. Unlike organics, which is certified by the EU and national bodies, biodynamics is certified by a number of different bodies that set their own standards (which are always stricter than organics in areas like copper use; usually half that of organics). Today the leading certification agency for biodynamic wines is Demeter, a global non-profit organization that gave its first biodynamic label to a Mexican coffee farm in 1928.

The other main agency in France is Biodyvin, which is only concerned with wine. Together they represent around 500 estates in France alone, and the practice is growing at around 10% per year, helped by the influence of leading consultants such as Pierre Masson and soil microbiologists Claude and Lydia Bourguignon, both of whom work with many of the producers in the book.

An estate must be organic before applying for biodynamic certification (or can begin both at the same time, but the process takes a little longer). The usual certification time is three years, although this varies slightly per country and certification body, see p.256 for contacts in various countries.

Both Demeter and Biodyvin ask vineyards to use a minimum of BD (biodynamic) preparation 500 (or 500P), 501 plus a biodynamic compost. There is a separate process for winery certification for Demeter in the US that involves a 14-step process that ensures full traceability and minimum intervention; SO_2 is allowed but at lower levels than organics.

Above: Cow horns are buried throughout the vineyard. Opposite: Château Climens in Sauternes grows and prepares all its biodynamic treatments.

Soucis
Camomille
Pissenlit
Achillée
millefeuille

Natural Wine

The term Natural Wine has become more and more widespread over recent years, cropping up on wine lists the world over. In Paris alone there are dozens of wine bars and restaurants that serve nothing else. It's tough to tell you exactly what it is, however, as, unlike organic or biodynamics, there is no agreed-upon rule or certification body governing it.

In theory, natural wine is made without chemical fertilizers or pesticides in the field, and no filtration, added yeast, enzymes and very little added sulphur in the wine. But because there are no official rules governing natural wines, there is nothing in theory to stop grapes being grown using conventional farming methods and then the wine being categorized as natural because no additives are used in the cellar.

Opposite: The philosophy behind natural wine is minimal intervention at every step of the winemaking process.

ANGIOLINO
MAULE
MASIERI
Opok
MARIA & SEPP

To be honest, I thought long and hard about including natural wines in this book. I learned much of my wine knowledge here in Bordeaux at the school of oenology, from legendary winemakers like Denis Dubourdieu and Jean Claude Berrouet. They are scientists, oenologists, researchers; they would say that winemaking is a process of not controlling but guiding nature, and would point out that grapes left without any intervention might ferment their sugars to alcohol but would quickly turn to vinegar. I've tried several natural wines that have tasted unstable, as this lack of intervention means the risk of bacterial infection and oxidization, among other things. But I've tried many others that are wonderful, and the theory of minimum intervention at all steps of the process is one championed by each and every winemaker in here. Essentially, the best natural winemakers share exactly the same philosophy as the best organic and biodynamic winemakers, and aim to deliver the most accurate reflection of their vineyard to your glass.

PERFECT SENSE

I was helped along the way by one of my favourite winemakers, Maria José López de Heredia in Rioja (see p. 134). Her wines are astonishing, and a visit to her winery is like a step back in time. She ferments in large open wooden vats, uses only indigenous yeasts for the first fermentation and then leaves the resulting wine to finish its secondary 'softening' (malolactic) fermentation in its own sweet time (some winemakers would add enzymes or heat up the room at this point to speed things along).

Above: Maria José López de Heredia works in ways that her father and grandfather would easily recognize.

Maria José goes even further, as she has no temperature control in the vat room and has been known to simply leave the windows open when a little extra air is needed. Basically she approaches the whole process exactly as her father and grandfather did before her. I have loved and admired the López de Heredia wines for years – and it made perfect sense when she said, 'We have been making natural wines for over 100 years'. It made me realize that the label is less important than the approach of the individual producer, and their commitment to drawing back the veil between the vineyard and the final product in the glass. There are plenty of conventionally made wines that lack any feeling of soul or connection to their vineyards, which is maybe the biggest disappointment you can find when you open a bottle. Natural wines are the antithesis of this.

But, choosing who to include in the book was still difficult. I am hoping to share the world's best artisan wines, and I remain sceptical of natural for natural's sake. So I had to come up with some rules for inclusion.

For this, I went to the Bourguignons, as the true experts in the field. They advised that organic or biodynamic rules should be applied in the vineyard, and that during winemaking there could be no additives to compensate for grapes that are either too low or too high in acidity, no added sugar to increase alcohol, no added yeasts to help control fermentations, no added enzymes to help with malolactic fermentation, but very low amounts of added sulphur could be used for stability. Natural wines in this book are from winemakers who have proved quality consistency.

Above: Working by hand is key at Raventós i Blanc in Penedès, Spain.

Orange Wine

Orange wines – or amber wines as they are also known – get a hard time for being even more niche than natural wines, but, in fact, use one of the oldest known methods of production, which gives built-in protection to the wines. They are made by leaving white grapes in contact with their skins (and sometimes seeds and stalks) for an extended period of time, unlike typical white winemaking, which separates the juice from its skin very early in the process. To put it another way, orange wines are white wines made like they are red.

Think of them, I would say, like a lambic beer, where wild yeasts and bacteria are left to do their thing and the fermentation process can take weeks, months or even years in some cases.

Opposite: Orange wines are some of the most food-friendly wines you can find.

AGENO
2011
LA STOPPA

The resulting wines will have a deeper golden colour than a normal white, and a deeper intensity of flavour and aromatics on the nose and palate. They often have the kind of tannic grip and structure that you would expect to find in reds.

This style of winemaking dates back for centuries in areas such as Georgia, typified by the *qvevri* wines made in egg-shaped earthenware vessels (since 2013 this method has been on the UNESCO World Heritage list of intangible cultural heritages). It is also used in northern Italy, particularly in Fruili by legendary producers like Radikon and Gravner.

All of this makes them extremely food-friendly, yet challenging if you open them at the same time as traditional whites because they are so different. My advice is to give them a few minutes in the glass to allow the mandarin and soft marmalade notes to reveal themselves, and be ready for a more structured feel in the mouth.

The orange wines I've chosen for this book almost all come from already-certified biodynamic estates, like Meinklang in Austria. And for food-pairing suggestions, I've enjoyed chatting with journalist and orange wine specialist Simon Woolf, who says their flexibility is what really makes them so interesting and worth thinking about when choosing wine to go with a meal.

'Skin-contact gives you the structure of a red wine with the freshness and energy of a white – something that lends huge versatility to food pairings. I've had good results with everything from fatty charcuterie to spicy Asian cuisine.'

Above: Egg-shaped cement vats (Meinklang, Austria) retain grape intensity and character during winemaking. Opposite: Grüner Veltliner at Weingut Hirsch, Austria.

Other Sustainable Practices

BIODIVERSITY, PERMACULTURE & HEALTHY SOILS

It's all very well to admire winemaking regions that extend for mile after mile with perfectly clipped vines. But the best farmers understand that monoculture is not a good idea for long-term sustainability of their vineyards or the environment. Instead, they encourage a healthy ecosystem and mix of plant (flora) and animal (fauna) life in the vineyard. The more genetically diverse the interactions are between flora and fauna, the more sustainable the vineyard is likely to be. This might mean through attracting beneficial insects and birds that will reduce the need for pesticides or through plantings of hedgrows and shrubs to create shade that can lower vineyard temperatures in hot climates or provide wind shelter to minimize drift and waste from any vineyard sprays. Permaculture is a further extension of this, where agricultural systems are created that can be sustained indefinitely by recreating natural ecosytems.

I asked myself many times what the single most important criteria in deciding which wines to include in this book should be. I kept coming back to the words of Claude and Lydia Bourguignon, who have been so wonderfully helpful and patient as I have asked them questions along the way. The focus of much of their work lies in showing how soils have to be alive to produce the best grapes, and that biodiversity and biodynamic practices in particular are the best way to encourage the micro-organisms that ensure soil health. In fact, their advice to me was, 'Pick up a handful of soil when visiting a vineyard and smell it. Healthy soil full of life smells good.'

All organic and biodynamic producers understand the importance of this – to the extent that most certification programmes make biodiversity a condition of the accreditation. Demeter, for example, insists that all certified farms must have a minimum of 10% of total land base set aside as a biodiversity reserve. We considered a symbol in the book for wineries that are particularly strong in this area – Spring Mountain Vineyards in Napa and Domaine Comte Abbatucci in Corsica spring to mind, but in the end I decided that all of the winemakers chosen have shown commitment to biodiversity, and deserve recognition for it.

CARBON NEUTRAL/OFF-THE-GRID

Besides ensuring biodiversity, an increasing number of wineries are looking at ways to reduce their carbon footprint and other environmental

Opposite: Sustainable winemaking is all about preserving the vineyard for future generations.

impacts by becoming independent of fossil-fuel energy resources. This might be by installing alternative sources of power, from solar energy and geothermal heating systems for directly powering their wineries, or by initiatives such as planting drought-tolerant shrubs that don't need any water in their maintenance and so reduce the winery's overall useage of water and any machinery needed to distribute it. Wineries are well placed for alternative power sources because a lot of their raw matericals can be recycled – stems and pips, for example, that remain behind after destemming and crushing the grapes can be used an an organic fertilizer. Wineries can also use biodegradable paper for their cartons, and use lightweight bottles that use less glass and are lighter for packaging and delivery.

In the US, the voluntary certification programme LEED (Leadership in Energy and Environmental Design) offers a rating system for energy-efficient buildings, which is one of several international standards. Other programmes include Wineries for Climate Protection (WfCP) that was developed by the Spanish Wine Federation to show a winery is commited to the reduction of greenhouse gases, water and water managements and renewable energy.

Individual areas also have progammes to encourage responsible winemaking, such as Napa Green in California, a voluntary certification programme to show how a winery takes concrete measures to conserve resources, enhance efficiency and work with their employees to ensure sustainability. These are all excellent programmes, but are not always connected to farming organically or biodynamically, and therefore are not specifically highlighted in this book.

INDIGENOUS GRAPES

An indigenous grape is another name for a native grape – meaning one that is long-established in a particular area and has proved well-suited to its soils and climate. I'm thinking of grapes like Perricone in Sicily or Plavac Mali in Croatia, or Aglianico in Puglia, southern Italy. They might have been former wild grapes that were tamed along the way, or come from natural crosses with other varieties in the vineyard, or just have been brought to a region so long ago that they adapted and developed their own specific character. They bring diversity and personality to a wine region, and are often talked about in contrast to the big international varieties like Cabernet Sauvignon and Merlot that together hold the two top spots as the world's most planted red varieties.

There is often a link between the use of indigenous grapes and sustainability, because they are naturally suited to the local environment, and so often have a better resistance to local conditions, such as regularly hot summers or cold winters – all of which means a better natural balance in the grape and less need for intervention in the cellar, such as added acidity or sugar.

Indigenous grapes are frequently little-known and so winemakers choose to plant the big-name grapes to attract higher attention and prices. We can help by making buying choices that support indigenous varieties – and by remembering that when it comes to food matching, native grapes are often best (as sommelier Jeff Harding says, 'What grows together, goes together').

SULPHITES

Sulphur (SO_2) is a naturally occurring element and is harmless to most people. It is permitted in organic vineyards as a non-toxic fungicide. Grapes themselves also produce sulphites as a natural by-product of fermentation, so it is impossible to have a truly sulphur-free wine. Instead you will find 'no added sulphur' on certain labels. Added during wine production or bottling, the compound SO_2 protects against oxidation and microbes, keeping wine fresh, stable and free of flaws throughout shipping and non-refrigerated storage.

In the EU, rules for organic wine allow a maximum of 100 milligrams per litre (mpl, also expressed as ppm) of sulphur for red wines, compared to 150ppm for conventional reds. For white wines

Opposite: A healthy soil is crucial for long-term sustainability.

this rises to 150ppm (compared to 200ppm). Canada follows the same amounts. The US allows lower levels of only 10ppm if they are labelled organic, and up to 100ppm if their label includes the wording 'wines from organic grapes' (compared to 350ppm for conventional wines, for reds and whites).

For biodynamic wines certified by Demeter, sulphur additions of up to 70ppm is allowed for red wines and 90ppm for whites. Natural wines, as defined by the Bourguignons and used as standard for this book, allow 30ppm for reds and 40ppm for whites. In Australia, organic and biodynamic wines allow up to 120ppm (compared to 250ppm for conventional wines, for red and whites).

For the purpose of comparison, a packet of dried apricots usually contain somewhere between 500 and 2000ppm.

And Finally...

The wines here are all examples of best practice. But no system is bullet-proof. There are always instances, in some vintages, when winemakers have no option but to use traditional treatments. For the purposes of this book, if this happened in recent years as a genuine last resort and not a regular occurrence, we gave them a pass. By the same token, not everyone in the book is 100% organic or biodynamic across all of their vineyards, which will annoy plenty of fully signed-up and certified producers I know. But, I wanted to draw attention to a few figureheads who are helping to spread the word, including Seña in Chile, Torres in Spain and Louis Roederer in France. Purisits will say they shouldn't be in here, but I believe their influence on the acceptance of these farming methods and the benefits it brings to the taste of wine makes them absolutely worth celebrating. To make this clear, these wineries will display the symbol Low Intervention (see p.12).

Miguel Torres Senior is one of the wine world's most outspoken champions of green winemaking practices, and he is honest about the challenges. 'As viticulturalists, we can't escape the fact we are polluting. The biggest source of emissions in vineyards is from tractors, something that almost all vineyards use whatever system of agriculture they follow. And even if we get close to being carbon neutral in the vineyard, then we have to accept that an inevitable byproduct of turning the grapes into wine is the production of carbon dioxide.'

The point he is making is that every system will have its detractors. But that's okay, because all of these approaches are inevitably part of an overall commitment to sustainability and quality. They take time, effort and money, and require a deeper understanding of the vineyard. They mean that the winemakers who practice them have to stay close to their vines, and the resulting wines taste better because of it. The fact that we can also feel good about enjoying them because they follow the philosophies celebrated here is simply an added bonus.

Above: Chardonnay at Louis Roederer, Champagne.
Opposite: Elisabetta Foradori is one of the leaders in biodynamic farming in Italian wine.

How To Choose Wine

So, you're ready to start choosing wines by their feel-good factor, and by the love, skill and commitment to craftsmanship that the winemakers put in each and every bottle. But where do you start?

Certification labels on bottles is of course the most reliable guarantee, particularly because producers are not allowed to write on their label that they are organic or biodynamic unless they have received official proof, even if they follow the methods in the vineyard. So to be certain of what you are buying, look for the symbols Ecocert, Demeter, Biodyvin, CCOF, USDA Organic and others on the label (see also p.256).

Having said that, it's not always easy to identify organic and biodynamic wines, because not all grapes grown organically or biodynamically are certified, and not all wines made from certified organic or biodynamic grapes display the fact on the label (or if you are picking from a wine list, then the individual bar or restaurant possibly won't spell this out). I can't tell you how many of my wine friends were surprised to learn that many of the most iconic names included in this book are in fact working organically or biodynamically. They just didn't know, often because the wineries are sought after enough without advertising; yet another string to their bow (yes, Cristal Rosé, Inglenook, Henschke and Domaine de la Romanée Conti, I'm looking at you).

And then there are natural and orange wines, which, as I mentioned, don't yet have a certification programme available to them (although apparently the EU is working on one for natural wines). It's always worth asking sommeliers and retailers for advice in this case, many of whom will know much more about the back stories involved in the wines than are listed on the back label, and most importantly will be able to direct you towards

smaller names and exciting discoveries that you shouldn't miss. Almost inevitably, these types of winemakers who are pushing at the boundaries of viticulture and often working on small-batch, handmade wines don't have big distribution deals or marketing budgets that put them on supermarket shelves or on billboards around town. So, it's also worth going to specalist shops and wine bars that celebrate and stock them. Think of this as another way you can support native grapes and small-scale wine business in all parts of the spectrum.

I really hope this book gets you excited about trying out a ton of different wine styles, and to think of discovering these producers just as you would a new ingredient, or style of food. I hope you will start to think of them as just another essential element to a good meal, and within the bigger spectrum of what makes eating and drinking such an enjoyable thing to share with friends. To encourage you further in this, I have asked the brilliant Jules Aron, known as the Healthy Bartender, to suggest some wine cocktails on p.74, which are especially suited to these artisan producers who focus on delivering clean, crisp flavours.

As Jules says, 'Almost any flavour can be highlighted by wine, and the complexity of flavours makes the possibilities for using them as ingredients in cocktails practically limitless.'

To give further inspiration, we have put together a directory of bars, restaurants and stores in many key cities around the world that specialize in the types of wines featured. Although it is by no means exhaustive, it includes many of our favourite places, and those associated with the sommeliers who have provided such brilliant food pairing suggestions throughout the book. To find this, turn to p.251.

Above: Terroirs Wine Bar in London is one of many places to try artisan wines.

01

Sparkling & Fresh, Crisp Whites

Philippe Jamesse

Sommelier at Les Crayères, Reims, France

My father is a pastry chef, and my plan was to follow him into that career but my mother made a mistake on the application form to the culinary college, and I ended up enrolled in the sommelier class instead. I've been in love with wine ever since – of course Champagne, a region where I have always lived and worked, but also many other styles of sparkling and still wines.

I love how Champagne is seen as perfect for celebration, with a formidable ability to sharpen the appetite without wearing out your taste-buds. But it is also a gastronomic wine that can match foods through an entire meal. It is capable of ageing, of evolving over time, deepening in expression and therefore able to match both subtle and intense flavours. The effervesence gives an extra dimension, acting as a justaposition between flavour and texture, both intensifying and liberating each one. Organic and biodynamic viticulture increases this ability, emphasizing the chalky nature of the terroir and therefore the potential of the wines that come from it.

When food matching with all sparkling wines, the aim is to think about the overall structure, balance and character of the wine, and how these match with the ingredients and the style of cooking. When tasting, consider the spinal cord of the wine, its axis through the glass. This is what conveys its purity, and will help you decide whether to pair with a single ingredient, or a more complex array, and how to cook them.

In the wine, think about:

how strong the bubbles are, and their size;
the density of the wine;
the strength and style of its fruits;
the weight of the mouthfeel, whether it has grip and tannic hold;
the intensity, type and style of aromas on display.

Broadly speaking, these six elements will lead to three main styles: firstly a sparkling wine dominated by the primary aromas of fresh fruit and flowers, or secondy one that displays aromatics that bring in some character from ageing in barrels, for example, or richness from the yeast lees, or thirdly one where complexity and depth comes from ageing.

For the first style of light, fresh, primary flavours, for both sparkling and still wines, I would suggest carpaccios, tartares or a range of seafood dishes that are lightly cooked or not cooked at all, and that emphasize the primary nature of the aromatics and flavours.

We do a lot of work with flowers and raw vegetables in our cuisine at Les Crayères, and these work perfectly with the first category of wines. Essentially, simple ingredients served as fresh as possible.

As the textures and flavours of the wines deepen, I suggest strengthening and deepening the flavours of the accompanying foods. For Champagnes and sparkling wines that have some oak-ageing character, we will add depth to the sauces, lengthen the cooking time, perhaps integrate another element to the sauce, such as lightly sautéed mushrooms or a touch of cream. The wines can take this extra weight and the two will combine to increase the sensation of each one.

For the third category, which may be vintage Champagnes or wines that have developed bottle complexity over ageing, you can push the flavours even further. Intensity in the wine or Champagne allows for more intense methods of cooking, using higher heat, for example, or charcoaling, and the use of richer flavours in the food. Black truffles can be a wonderful match.

Philippe's Wine Picks

ANDRÉ OSTERTAG PINOT GRIS AND RIESLING, ALSACE, FRANCE

BENOÎT MARGUET CHAMPAGNE, FRANCE

BENOÎT LAHAYE CHAMPAGNE, FRANCE

CHAMPAGNE FLEURY PÈRE ET FILS, FRANCE

DOMAINE VACHERON SANCERRE, LOIRE, FRANCE

Sparkling Wine

It is hard to beat sparkling wines for flexibility when it comes to food matching, because they have such an array of different weights and tastes depending on their origin and style.

LA GARAGISTA GRACE AND FAVOUR PÉTILLANT NATURAL WHITE VERMONT 2014, USA

WWW.LAGARAGISTA.COM/WINES

A Pét-Nat rather than full-on sparkling wine, which means the wine was bottled before finishing its first fermentation, and so the bubbles are from CO_2 produced as a natural byproduct of fermentation. This is grown in Vermont on hybrid American vines called La Crescent, making the delicious results seem rather incredible. The people to thank are Deirdre Heekin and Caleb Barber and a lot of backbreaking work on their farm, a place where it's not at all unusual to have snow on the ground in April or May (they aren't far from Killington ski resort). Deirdre has written several books on biodynamics, including the brilliant *An Unlikely Vineyard* that I highly recommend. The wine is fermented in glass demijohns, with no added sugar or sulphur. It has beautiful orange blossom, apricot and fresh lemon flavours, with a crab-apple finish that tightens to an attractively tangy point just to remind you that this is not your usual glass of wine. And call me crazy, but my favourite match for this is an apricot tart.

ANCRE HILLS ESTATES WALES BLANC DE BLANCS 2009, UK

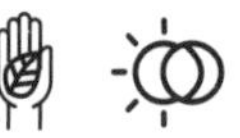

WWW.ANCREHILLESTATES.CO.UK

Full marks to Roger and Joy Morris for organic and biodynamic farming in the Welsh climate. Another successful biodynamic estate in the UK to look out for is Sedlescombe in East Sussex. The 2008 vintage of Ancre Hills was voted best sparkling wine in the world at a well-regarded Italian competition, and this follow-up vintage is thoroughly enjoyable. A 100% Chardonnay, the nose is focused, bright, aromatic, with gentle brioche notes; a beautifully dry citrus grip is set off by Granny Smith apples and soft elderflower. This makes a delicious glass of lunchtime fizz with goat's cheese toast. Demeter-certified.

ALBET I NOYA DO CLASSIC PENEDÈS BRUT NATURAL NOSODO RESERVA 2013, SPAIN

WWW.ALBETINOYA.CAT/ENG

Classic Penedès is a fully organic appellation, created in 2016, driven by this estate. Made in the traditional method with second fermentation in bottle, this is gulpable stuff. A blend of Xarel.lo, Chardonnay, Macabeu and Parellada grapes, this aged for 18 months in bottle before it is disgorged by hand with a tiny 6.5g/l of residual sugar. Lovely pale buttercup in colour, it is fresh and elegant, focused and floral with structured apple blossom and white pear fruit. It is totally unfussy so very food friendly, with dessert at the end of a meal or for cracking open with friends, a plate of Spanish olives and hams as an apéritif.

GRAMONA CAVA III LUSTROS RESERVA 2009, SPAIN

WWW.GRAMONA.COM

Rich in style, this has the finesse and depth you expect from the best sparkling wines. It has lemongrass, citrus, clear fresh acidity without being biting, with an undertone of grilled almonds and soft vanilla spice: a reassuringly impressive wine from Gramona and recognized as being at the top of the tree in Cava. Today it is run by Jaume and Xavier Gramona, the fifth generation of a family that began producing these wines back in 1881. Increasingly recognized for not just farming organically but also for serious and sustained efforts to encourage biodiversity and creating its own animal and plant composts that are now an essential part of their vineyard. This particular wine is 100% from the local Xarel.lo grape and is Gramona's first Demeter-certified bottling. Fermented in 300-litre French oak barrels, a mix of new and one or two years old, the wines are aged on the lees for three months.

Above: Gramona is leading the new generation of Cava producers.

Pepe Raventós
Raventós i Blanc

PL. DEL ROURE, S/N
08770 SANT SADURNÍ D'ANOIA, SPAIN
WWW.RAVENTOS.COM

I didn't actually realize who Pepe was when I started chatting to him – just that he spoke perfect English after living in New York for years, and that he was a brilliant advocate for terroir-driven Spanish wines. Also that his farm near Barcelona sounded like heaven.

When I tasted his wines it just added to the sense that this is someone who is really changing the system one step at a time. Besides getting to understand the consumers in the US market, Pepe has made wines in St-Émilion, St-Aubin, Pouilly Fumé (with the legendary Didier Dagenau), Alsace and the Nahe in Germany.

In 2012 he made headlines in Spain by leaving DO Cava to launch Conca del Ríu Anoia DO – the name of a tiny geographic area where his vines (and idyllic farm) are found. The quality charter for winemaking is tough – must be 80% estate-grown fruit, only organic or biodynamic cultivation and only indigenous varieties.

'I wanted to make wines that were an honest reflection of a specific piece of land and a specific climate. Being in New York made me even more convinced of that,' he told me. 'The whole family [he has four children] adored living there, but it underlined what we really have is an identity here in this land, and something to offer wine-lovers.'

> 'There is just so much pleasure to be had from wines that get close to the raw spirit of the land.'
> *Pepe Raventós*

RAVENTÓS i BLANC
VITICULTORS DES DE 1497
CONCA DEL RIU ANOIA
DE LA FINCA
VINYA DELS FÒSSILS

RAVENTÓS I BLANC DE LA FINCA 2013, SPAIN

Owner Pepe Raventós has recently moved back from New York full time to his family estate (he's the 20th generation), Raventós i Blanc. Located around 48 km (30 miles) from Barcelona, this wine has lovely fresh flavours of citrus and soft crab apples; medium-weight and ever so elegant in style, with touches of crème patisserie rather than the biscuity notes sometimes found in Champagne. Mouthwatering salinity gives a kick to the finish, providing a lovely match with oysters. Adding to the interest is that the wine is made from grape varieties indigenous to the region, with a blend of Xarel.lo, Macabeo and Parellada; it is aged for three years and bottled with zero dosage. Demeter-certified.

Ton Mata

Recaredo Mata Casanovas S.A.

TAMARIT 10, APARTAT DE CORREUS 15,
08770 SANT SADURNÍ D'ANOIA (BARCELONA), SPAIN
WWW.RECAREDO.COM

I first got to know Ton Mata in the heat of late June sunshine during a weekend in Spain, which had brought together the new wave of the country's winemakers. I can still remember walking over to try a wine that I heard so much about, and being blown away by this bright, energetic man who has slowly but surely been reinventing Cava.

Forget all that you know about standardized, big-business Cava when you meet Ton. He's the grandson of the founder of Recaredo, and is a restlessly moving coil of energy, softened by his wide smile and just-perfectly-oversized black-rimmed glasses.

The biodynamic vineyards are not the only place where Ton is rewriting the rules. For a start, he bottles all the wine according to the year that it was harvested, meaning only vintage Cavas are made – entirely from Penedès' native but underused grape Xarel.lo, grown without any irrigation even in the heat of a Spanish summer. The resulting wines are aged for a minimum of five years on the yeast lees to encourage rich, smoky flavours. Every step of the process is done by hand, and no sugar is added to the final wine at the moment of bottling, making the wines Brut Nature – emphasizing even more clearly that this is terroir-driven Cava is of the highest order.

'Recaredo for me means to be free. Doing what we want for the creation of wines that we like.'
Ton Mata

RECAREDO BRUT DE BRUT FINCA SERRAL DEL VELL BRUT NATURE GRAN RESERVA CAVA 2006, SPAIN

The first ever Demeter-certified estate in the Penedès region (where 90% of Cava is produced); receiving the official nod back in 2010 after four years of conversion. Here smoky, nutty notes melt into a ripe citrus body with a crystalline finish. This pairs well with a wide range of dishes and easily stands up to sauces because of its good palate weight. From a blend of local grapes, 53% Xarel.lo and 47% Macabeu, this is a lovely reflection of the abundant sunshine of the Cava region, while still delivering a clear sense of energy and finesse.

CHAMPAGNE AGRAPART GRAND CRU BLANC DE BLANC TERROIRS EXTRA BRUT, FRANCE

WWW.CHAMPAGNE-AGRAPART.COM

Champagne Agrapart is not certified organic, but never uses chemical pesticides or weedkillers; it uses homeopathic treatments for any vineyard diseases, ploughs with horses, ferments using natural yeasts and bottles without fining, filtration or cold stabilization. This is a beautifully fine, elegant and floral-scented Champagne that is made with grapes from 25 to 40-year-old vines in the Côte des Blancs *grand crus* of Avize, Cramant, Oger and Oiry. It is partly aged in large-sized wooden casks then spends just under four years in bottle before release. It has low dosage of 5g/l, and is certainly easy to drink on its own, but would be utterly delicious with a few just-warmed mini crab cakes.

CHAMPAGNE DE SOUSA MYCORHIZE GRAND CRU EXTRA BRUT NV, FRANCE

WWW.CHAMPAGNEDESOUSA.COM/EN

I can't recommend highly enough getting hold of this 100% Chardonnay fizz – or any of the Erick de Sousa range of terroir-driven Champagnes. This cuvée is named after the association between mushrooms and the vine roots, which points to a healthy soil; in this case a plot of old *grand cru* vines farmed biodynamically since 1999, ploughed by horses and fermented in barrel using only indigenous yeasts. With great intensity on the nose, its structure is light, delicate, with egg-shell finesse and a soft-lace texture that builds through the mid-palate along with notes of buttery patisserie. Only 3g/l of sugar is added at disgorgement, meaning it is extremely dry. Of all the food pairings I have tried this with, salmon sashimi has been the best. Demeter-certified.

CHAMPAGNE FLEURY PÈRE ET FILS EXTRA BRUT SONATE 2011, FRANCE

WWW.CHAMPAGNE-FLEURY.FR

The original green Champagne producer – organic since 1970 and biodynamic since 1989. This particular bottling was created by Jean-Pierre Fleury and his son Jean-Sébastien to celebrate their 20th year of biodynamics. Intense flavour from the largely Pinot Noir grapes (76%, with the rest Chardonnay), great focus and finesse, toasted almonds, strong citrus, persistent bubbles that kick up their heels long into the night. A total crowd pleaser every time I have opened it for friends, as are all the Fleury Champagnes. This calls for the French version of tapas – Comté or Gruyère *gougères*. Demeter-certified.

Previous and opposite: The red Pinot Noir and Pinot Meunier grapes are regularly used in sparkling wines, as here at Champagne Fleury.

CHAMPAGNE FRANCK PASCAL RELIANCE BRUT NATURE

WWW.BLOGFRANCKPASCAL.OVER-BLOG.COM

Serious green credentials here, as this cuvée was served at the Cop21 Climate Change Conference in Paris in 2015. This is a lovely grower Champagne, with all bottlings fermented using wild yeasts, spontaneous malolactic fermentation, no temperature controls in the cellar; bottled unfined, unfiltered, with minimum sulphur (never higher that 35ppm). This is a blend that focuses on Pinot Meunier (60%) and Pinot Noir (25%) with a 15% lift of Chardonnay. There is texture and weight in the body, and the beautifully round mid-palate of lemon is pepped up with lime blossom finish, working even with main courses like pan-fried white fish in a rich buttery sauce. Ecocert-certified.

CHAMPAGNE FRANÇOISE BEDEL ENTRE CIEL & TERRE, FRANCE

WWW.CHAMPAGNE-BEDEL.FR

Françoise works with her son Vincent at this 8.4 ha estate that has been biodynamic since 1998, both in the vineyard and across all their bottlings. This is particularly lovely Champagne from vines grown on limestone and clay soils by a winemaker who really cares about what she is putting in the bottle. Lovely crisp and clean flavours, very much focused on minerality, make this wine a perfect match for a heaped plate of seafood. Ecocert-certified.

Above: Benoit Marguet ploughing his vineyards with horses.

Benoît Marguet

Champagne Marguet

1 PLACE BARANCOURT, 51150 AMBONNAY, FRANCE
WWW.CHAMPAGNE-MARGUET.FR

The fifth generation of winemakers in Champagne, Benoît Marguet worked with Paul Hobbs in Washington State in the US before returning to Champagne back in the late 90s. He became close to a few young growers who were working organically like David Léclapart, and became interested in trying it out for himself.

'I started trying to convince my parents to plough their soils and go fully organic, but there was some resistance at first,' Benoît says diplomatically. They should have been used to it, as it turns out he had done a school project in 1997 on Nicolas Joly (ah to go to school in France hey), but they were not convinced. He eventually took a four year break from the family estate to start his own merchant business, before returning in 2008 and continuing what he had started.

Today, certified both organically and biodynamically, Benoît ploughs the vines with horses (two of them – named Titan and Urban) and uses various preparations from herbal infusions to aromatherapy on his vines.

> 'I just try to take a global approach – everything is connected and in communication with each other, so my job is simply to clear the path.'
> *Benoit Marguet*

CHAMPAGNE MARGUET AMBONNAY GRAND CRU 2010, FRANCE

Benoît Marguet goes a little further than some in his farming methods – using herbal supplements and aromatherapy oils (think liberal use of essential oils such as lavender and citronella). Probably the best-known natural wine producer in Champagne, these wines have zero dosage and no added sulphur. This vintage *grand cru* blend is 58% Pinot Noir, 42% Chardonnay, disgorged in February 2016 after five years on the lees. Acidity is relatively high, with the emphasis on apple and citrus notes. Expect a clean, fresh glass of fizz, firm structure and lovely purity of flavour. Just add a plate of grilled almonds.

CHAMPAGNE LECLERC BRIANT CUVÉE LA CROISETTE BRUT, FRANCE

WWW.LECLERCBRIANT.FR/UK

Now owned by Americans Mark Nunnely and Denise Dupré, but run by the sixth generation of the Leclerc family, Ségolène Leclerc together with Frédéric Zeimett, Leclerc Briant has been organic since the 50s and experimenting with biodynamics since the 60s making them clear early adopters. They even produced their first Zero Dosage in 1970. Although most of the original vineyards went to Louis Roederer in 2012, they kept La Croisette. Frédéric – who is also a part-owner – came to the property via Moët & Chandon. With hazelnuts and almonds, saline touches on the finish give a counterbalance to the rich and round mid-palate; this wine just has a beautiful depth and focus. Zero dosage, and long on flavour. Demeter-certified.

CHAMPAGNE PASCAL DOQUET ARPÈGE PREMIER CRU BLANC DE BLANCS EXTRA BRUT, FRANCE

WWW.CHAMPAGNE-DOQUET.COM

Pascal Doquet is the head of a grouping of organic producers in Champagne, and has worked without chemicals since 1982; something that his parents at the time saw as a backwards step. 'I was the only one in the family to see the use of chemicals as a dead end,' he likes to say today, which is probably also why he is also a big believer in certification. 'It is so hard to work organically in Champagne and so easy to say that you are doing so.' Arpège is a blend of three Chardonnay plots across the Côte de Blancs (Vertus, Villeneuve and Le Mont-Aimé) aged for six months on lees and a blend of two vintages (2010 and 2011). A delicate, softly fragrant wine that is more minerally than fruity, yet is surprisingly rich and full after a few minutes in the glass. This stands up excellently to a spiced tuna carpaccio.

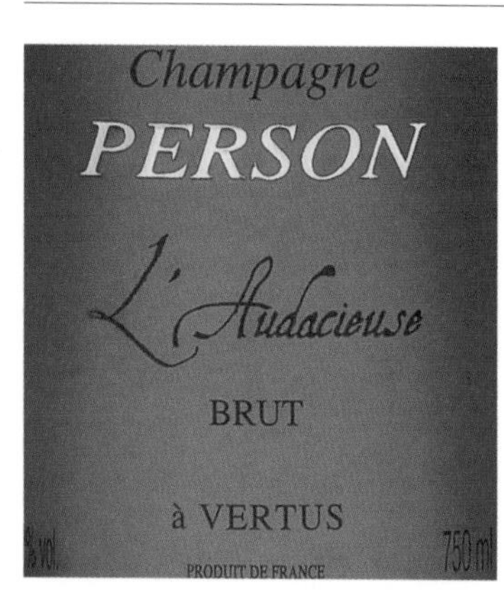

CHAMPAGNE PERSON L'AUDACIEUSE BRUT NV, FRANCE

WWW.CHAMPAGNE-PERSON.FR

A 100% Chardonnay Champagne from Dominique Person of Domaine Le Clos des Belvals. This is pure but rich, with great depth of white flowers and mandarin flavours and a slightly smoky kick that delivers an effortless sense of fun. Vinified in Burgundy barrels, it is a super-dry style with just 5g/l dosage. Dominique has only been farming biodynamically here since 2009. Keep food matches very simple here, to focus on these tight, delicious flavours – asparagus risotto perhaps.

DAVID LÉCLAPART L'ARTISTE EXTRA BRUT BLANC DE BLANCS PREMIER CRU, MONTAGNE DE REIMS 2007, FRANCE

10 RUE DE LA MAIRIE, 51380 TRÉPAIL

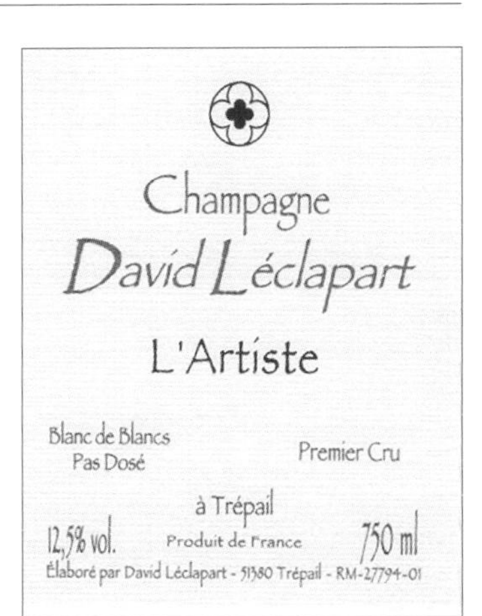

The grapes come from 30 to 50-year-old Chardonnay vines grown in the village of Trépail in the Montagne de Reims. The wines are always the product of a single harvest (so no reserve wines in this cellar), and things are kept equally focused for winemaking – no chaptilization, wild yeasts to start fermentation, no added enzymes for malolactic, no filtration, no added sugar. The Artiste bottling is aged 50% in vat and 50% in used oak barrels (usually from Domaine Leflaive in Burgundy) and kept for 40 months in bottle before disgorgement. You get toasted brioche in full effect here, beautifully complex with lovely chalk-tinged bubbles. A beautiful match to seared scallops. Demeter-certified.

DOMAINE DE LA TAILLE AUX LOUPS TRIPLE ZÉRO MONTLOUIS SUR LOIRE, FRANCE

WWW.JACKYBLOT.FR

One of the towering winemakers of the Loire, Jacky Blot's Triple Zéro is one of the insider sparkling wines of France. From 100% Chenin Blanc, all vines are 50-years-old+ and it's fermented with natural yeasts. The Triple Zéro of this wine's title comes from the fact that there is no sugar added to the grapes at fermentation (a technique used to add alcohol), nor before the second fermentation that gives the bubbles, and no dosage added at the moment of bottling. What you get is crisp, focused notes of citrus and white peach, as bone dry as you like. A perfect brunch choice with scrambled eggs and chives.

DOMAINE PIERRE FRICK CRÉMANT BLANC DE NOIR ZÉRO SULFITES 2011, FRANCE

WWW.PIERREFRICK.COM

Twelve generations of the Frick family have been making wine here, at an estate that has been organic since 1970 and biodynamic since 1981. They do a range of zero-added-sulphur still wines, but I really recommend this delicious 100% Pinot Noir Crémant. Get yourself a plate of smoked salmon and enjoy the slightly smoky notes of roasted sunflower seeds that just curl up out of the glass, deepening the gently ripe apricot and citrus fruits.

'So good with duck liver mousse, charred sourdough, red wine gelée.'
Caleb Ganzer, Wine Director and General Manager, La Compagnie des Vins Surnaturels, New York

Jean-Baptiste Lecaillon
Louis Roederer

21, BOULEVARD LUNDY, 51722 REIMS, FRANCE
WWW.LOUIS-ROEDERER.COM

A few years ago, just after the hillsides and cellars of Champagne received their UNESCO World Heritage status, Jean-Batiste Lecaillon drove me out to the vineyards where the grapes for the Louis Roederer Brut Nature Champagne were grown. We walked through the vines and he talked me through the focus on biodynamics that has been growing ever since he made the decision to stop all herbicides back in 1999 and began the conversion to biodynamics from 2001.

This is a man who has all the effortless charm and brilliance that you might expect from the vice president of a Champagne house, and yet he is a constant surprise. For a start, it is hugely rare in the Champagne region to have a *chef de cave* – Jean-Baptiste is also the winemaker responsible for the blending – who is in charge of day-to-day strategic decisions; he works alongside owner Frédéric Rouzaud. So when the trials of biodynamics showed the potential for deepening the vitality of not just the soils but the wine, Jean-Baptiste decided the best place to start was at the top – with Cristal.

'It makes sense that we would want to give voice and expression to our best soils,' he said simply.

They are still not all the way there, but of the 240 ha owned by Louis Roederer, over 100 are farmed biodynamically.

'We're not looking for certification,' Jean-Baptiste says today, 'we do it instead for what we believe is best for the land and the quality of the wine. We found less of a taste difference in the trials of organically farmed vines, so we concentrated on biodynamics. It is also a philosophy that makes sense to us – it allows you to ask greater and wider questions about the ecosystem and our place within it. Right now we are considering stopping ploughing and composting altogether on these soils, and working towards a fully stable long-term system with the absolute minimum intervention. When you begin to work this way; you are always conscious of how you can go further.'
Jean-Baptiste Lecaillon

Above right: Today over 100 ha of Louis Roederer vines are farmed biodynamically, including all grapes for Champagne Cristal.

LOUIS ROEDERER CHAMPAGNE CRISTAL ROSÉ 2007, FRANCE ✸ ✸ ✸

The first 100% biodynamic cuvée from Louis Roederer's iconic fizz came with this 2007 Rosé, and the entire company became the largest biodynamic producer in Champagne in 2012 with the purchase of Champagne Leclerc Briant (this part is certified Demeter). Unusually for such a large and high-profile house, today over 40% of the 240 ha are farmed biodynamically. This includes every single grape that goes into Cristal, which hopes to be certified by 2020. This Champagne more than lives up to the high expectations we have of it. Sweetly delicate redcurrant fruits, from a blend of 56% Pinot Noir, 44% Chardonnay, produced by the *saignée* process of draining the juice from the skins after a few hours, when it is just lightly coloured. Silky, concentrated layers of rose petal, red fruits, blood orange and a touch of salted almonds. It is so light and elegant that it floats above ground, and yet the flavours keep on coming. It should age easily for a couple of decades. Food-wise, keep things simple and fresh, maybe a tuna sashimi.

Anselme Selosse
Jacques Selosse

HÔTEL RESTAURANT LES AVIZÉS
59, RUE DE CRAMANT, 51190 AVIZE, FRANCE
WWW.SELOSSE-LESAVISES.COM

To say that Anselme Selosse causes controversy is something of an understatement. He stubbornly sticks to the idea of single-vineyard terroir-led Champagnes, and even more stubbornly to a style of winemaking that is entirely his own – indigenous yeasts, vinification in barrel followed by ageing on the lees for up to a year before bottling, low sulphur and even the use of a solera system for specific cuvées.

So we shouldn't be surprised that he applies the same principle to organics – he works this way but doesn't want to be certified.

'For me every form of life in our vineyard is part of the terroir. Like the importance of the water source for whisky, so the micro-organisms in our soil are essential to transmit the flavour of the earth. For me presenting a wine is presenting a certain place. It must be a reflection of that place. I have always admired science, but in my vineyards I see this as being the only way of working. I would lose that transmission of terroir if I altered the microflora of the soil.

'At first some of my neighbours told me that my 19th century techniques would never revolutionize anything, but I am very proud today that many young winemakers and even big houses in Champagne are moving towards these methods of production. A strong motivating factor is the need to keep the name of Champagne as iconic around the world as it is today.'

'We winemakers are the witness and caretakers of our lands but we are not the creators. This is the only way to respect it.'
Anselme Selosse

JACQUES SELOSSE VO 'VERSION ORIGINALE' GRAND CRU BLANC DE BLANCS, FRANCE ✱ ✱ ✱

This is a producer who is close to the land, and who cares about reflecting its voice. The VO is a mix of three vintages of Grand Cru Avize, disgorged in July 2016. Rich, deep and round in colour, expect slightly oxidized, hazelnut notes. Not everyone likes Selosse because it is very much a 'vinous' Champagne – it tastes a lot like wine with a few added bubbles, so can be a surprise. It has stunning complexity and depth of flavour, and asks you to make a decision about whether you like it or not. Salinity is a key note that draws the poached-fruit flavours out, and in an evening with several different Champagnes, it was the Selosse that improved most in the glass. It has spent seven years in barrel and bottle and has just under 2g/l dosage. It could stand up to a truffle risotto or pan-fried duck because of its weight.

ERIC RODEZ CHAMPAGNE CUVÉE DES GRAYÈRES BRUT NV, FRANCE

WWW.EN.CHAMPAGNE-RODEZ.COM

A blend of 60% Pinot Noir and 40% Chardonnay, this is mouth-watering, with a floral, fresh style that emphazises the elderflower notes – a great match with a few slices of brie and saltine crackers. Today run by Eric and Martine Rodez with their son Mickael, Eric was an oenologist at Krug and also worked in Alsace, where he learned to appreciate the benefits of organic winemaking – and although he is not certified, he uses no pesticides, fertilizers or insecticides. This wine is a blend of five different vintages, with half aged in small oak barrels, half in tank with a light dosage of around 8g/l.

SAPIENCE CHAMPAGNE PREMIER CRU EXTRA BRUT 2007, FRANCE

WWW.CHAMPAGNE-SAPIENCE.COM

A popular choice around any table I have opened it for, this wine majors on salinity, oyster shells and slight hazelnut and sherry notes. Excellent as a match for food, this can take sturdy sauces, but is utterly delicious kept simple with oysters. Benoît Marguet partnered with a former *chef de cave* from Duval-Leroy to make this, with grapes from 35-year-old+ vines from across the region, all fermented in used white Bordeaux barrels. Ecocert-certified.

PERLAGE WINERY COL DI MANZA VALDOBBIADENE PROSECCO SUPERIORE DOCG 2016, ITALY

WWW.PERLAGEWINES.COM

This is made from 100% Glera grapes in the vineyards of Farra di Soligo, in the heart of the DOCG Prosecco, and is the first estate in the region to use certified organic grapes (they also follow biodynamic farming methods but are not certified). Secondary fermentation is in tank not bottle in the classic Charmat style, and has 17g/l of residual sugar that is barely discernable but gives a clear roundness to the mid-palate. Pale yellow in colour, apricot and fleshy white peach flavours, more depth than many Proseccos and dangerously easy to drink. Very flexible for food pairing, from a fresh buffalo mozzarella apéritif to a fruit dessert.

Fresh, Crisp Whites

Dry white wines, just like reds, have a staggering array of expressions, from crisp and light through to rich and round. The wines here are best with foods where the flavours are kept fresh and clean to reflect the sculpted purity of the wine. They are every bit as complex as the richer white wines in the next section, but focus on finesse and minerality.

OVUM TRADITION MEMORISTA RIESLING OREGON 2015, USA

WWW.OVUMWINES.COM

From the crack team of John House and Ksenija Kostic, this stunning Riesling is as gentle as you like and yet just seems to stretch out endlessly on your palate. Every wine that these guys produce is on a seriously small scale, so forgive me if it proves tough to get hold of – but it is so worth it when you do. Fermented in 1000-litre cement eggs and old barrels, expect orange rind, stone fruits, rain on slate and wafts of grilled smoke. The salinity on the finish is delicious, mouthwatering in the extreme. There is 35g/l of residual sugar here, but the bracing acidity balances it all out. A wine to relax with, grab yourself a glass and a fresh baguette and head out to the nearest garden for a picnic.

MATETIC CORRALILLO SAUVIGNON BLANC SAN ANTONIO 2016, CHILE

WWW.MATETIC.COM/EN

This unoaked Sauvignon Blanc is incredible value – ready to go, and full of bright citrus and soft apricot fruit. The vines, in the Rosario Valley in Chile, are known for making some of the country's best examples of this grape (and Pinot) because it gets those all-important ocean breezes from the Pacific. Matetic is just an amazing place, with floor to ceiling glass pretty much everywhere you look. Don't think too hard about this, just crack it open and pair it with a lovely goat's cheese salad. Certified organic and working biodynamically.

OPPOSITE, ABOVE: MATETIC OFFERS VALUE AND FLAVOUR IN CHILE.

CASA DE MOURAZ VINHO VERDE AIR 2015, PORTUGAL

WWW.CASADEMOURAZ.COM

Casa de Mouraz is all about getting right back to the basics of fruit, and here they ensure wild-yeast fermentation and low sulphur to focus attention on the raw materials. Loureiro (80%), Trajadura (10%) and Arinto (10%) are fermented in stainless-steel tanks at very low temperatures; there is a touch of residual sugar here (12.5%), which makes this wine a perfect match with raw fish – grab a plate of sushi or sashimi and you will be in heaven.

BIECHER & SCHAAL RIESLING ALSACE GRAND CRU ROSACKER 2015, FRANCE ✷ ✷

WWW.BIECHER-SCHAAL.COM

Julien Schaal is making brilliant wines out of Alsace and also Elgin in South Africa. Here he works his magic with one of Alsace's best-known *grand crus*, which owes the Rosacker name to the wild roses growing around the edge of the vineyard, and where limestone soils help flesh out a rich and dense Riesling. This is walking the tight rope; both easy to drink and yet thrillingly withdrawn. It is so dry you can feel it scratch down the back of your throat and yet it is so layered and complex it could keep you talking for hours. A dream with roast chicken but equally with a tandoori, it draws flavour out of any food.

Athénaïs de Béru
Château de Béru

32, GRANDE RUE, 89700 BÉRU, FRANCE
WWW.CHATEAUDEBERU.COM

It was a risky move, perhaps, to give up a stable job in finance to return to a family estate that had been rented out for the previous 20 years, but how often do you get the chance to create something that is both totally new and yet part of your family's heritage?

That, at least, was how Athénaïs de Beru felt when she moved from Paris back to her childhood home in Chablis, and gradually took over the vines, piece by piece, converting them to organics and then biodynamics as she went.

'I took a short professional conversion course in Beaune, but I really learned from friends and from tasting widely. I was always drawn to those who worked in biodynamics and to the energy in their wines. It was what got me excited about taking on this challenge.

'It is still incredible to me that I am now back in a tiny village of just sixty inhabitants and yet I am so connected to the world through my wine. I arrived full of passion for the project, but am not sure that I fully appreciated just what meaning the wines of Chablis have for people around the world, and how many people are truly knowledgeable about them.

> 'It still surprises and astonishes me that from one day to the next you can be in the vines pruning, almost in a meditative state, and yet at the same time be globally connected to wine-lovers. It is something perhaps that you only get in wine.'
> *Athénaïs de Beru*

CHÂTEAU DE BÉRU, CLOS BÉRU MONOPOLE CHABLIS 2013, FRANCE ✸ ✸

Clos Béru is a walled vineyard that dates back to the 13th century and lies on south-southwest facing slopes at 300m (984ft) in altitude. Fully ploughed by horse, this showcases the magic that can happen with Chardonnay in Chablis, where it is precise and tense on the attack and then slowly but surely builds in power until you are fully gripped by the subtle drama of slate and flint-infused lemon. This is not a wine that you want to let go of, and if you're anything like me you'll be buying another bottle as soon as the first one has gone.

DOMAINE STÉPHANE BERNAUDEAU LES ONGLÉS VIN DE FRANCE 2014, FRANCE

14 RUE DE L'ABONDANCE, 49540 MARTIGNÉ BRIAND

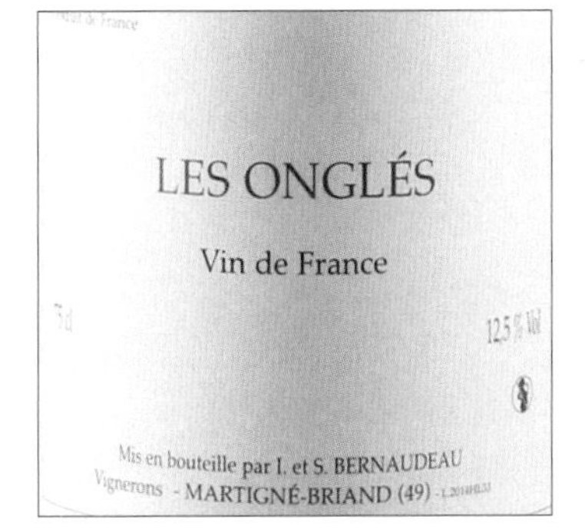

Old Chenin Blanc vines located between Angers and Saumur in Anjou, Loire, by Stéphane Bernaudeau, a darling of sommelier's worldwide, who is making a small range of wines. I had this particular bottling at Rouge Tomate restaurant in New York with three friends who work in wine. I was the only person who knew all three before the lunch, and this was the wine that made everyone sit up, get excited and start chatting to each other like old friends. Just a brilliant wine: ripe stone-fruit flavours, delicate but richly nuanced. Aged in old oak barrels, with natural yeasts and no fining or filtration. We had this with a selection of different antipasti and salads, and it was particularly good with a lentil salad.

'Try this wine with pork rillettes with diced fennel and apple.'
Caleb Ganzer, Wine Director and General Manager, La Compagnie des Vins Surnaturels, New York

---

DOMAINE VACHERON LES ROMAINS SANCERRE 2015, FRANCE ✱ ✱

1 RUE DU PUITS POULTON, 18300 SANCERRE

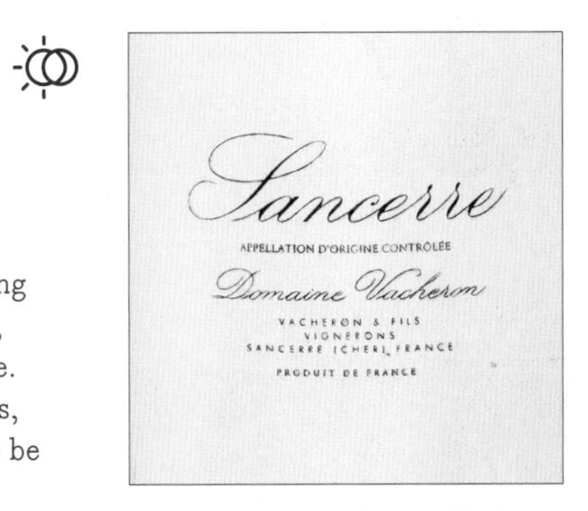

A small, low-key estate located right in the middle of Sancerre village; Cousins Jean-Dominique and Jean-Louis Vacheron save their impact for the wine. Razor-sharp Sancerre is my usual tasting note for their bottles, and once again they don't disappoint. Slate, citrus, mint and white pepper; taut and mineral, dancingly precise. 100% Sauvignon Blanc aged half in barely perceptible oak barrels, half in cement tanks. You just know a plate of scallops is going to be enhanced by a glass of this wine.

EMMANUEL HOUILLON/MAISON PIERRE OVERNOY ARBOIS PUPILLIN 2011, FRANCE

RUE ABBÉ GUICHARD, 39600 PUPILLIN

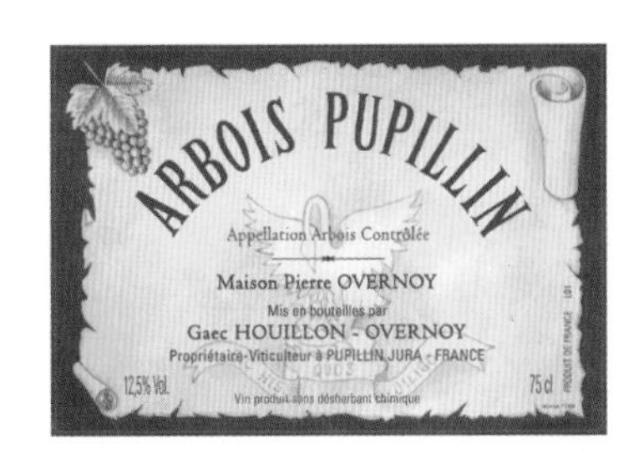

Maison Pierre Overnoy is a Jura estate, and sommelier's favourite. Pierre took over the estate from his father in the 60s, and took on Emmanual Houillon in 1990. He is now a partner and makes the wine with his brother Aurélien. Its organic credentials are right up there on the label, 'no chemical weedkillers', not the most enticing of tag-lines, but it certainly gets across how seriously these guys take natural wine (Pierre made his first wine with no added sulphur in 1984). Straw coloured, softly spicy, with a hint of smoky stubbornness, citrus acidity it is never prepared to permit easy descriptions. The smoky, almost salty side of this begs for a plate of oysters, but be sure not to choose anything that gets in the way of its beautifully pure fruits.

Julien Brocard
Domaine de la Boissonneuse

3 ROUTE DE CHABLIS - 89800 PREHY, FRANCE
WWW.BROCARD.FR/EN

I first met Julien Brocard around eight years ago, and his wines have been a regular feature in my fridge ever since. He's more welcome proof that scale doesn't have to get in the way of good intentions – the Brocard family are easily one of the most significant owners of biodynamic vineyards in Burgundy.

'When I came back to work with my father after finishing an engineering degree in Paris, I wanted to understand why so many winemakers were simply treating diseases in their vines without questioning why the symptoms had arrived, and whether chemicals were necessarily the best solution. The people who I spoke to who were similarly questioning this were biodynamic growers.

'I haven't found all the answers, and would need two or three lifetimes to do so. It can be hard to accept that you have to take a hit on yields for the first few years, and at times you have to go from failure to failure without losing your nerve. But once you remove the protective veil of synthetic treatments, you realize how much finesse and balance can be achieved, and how endless are the lessons we have to learn from nature'.
Julien Brocard

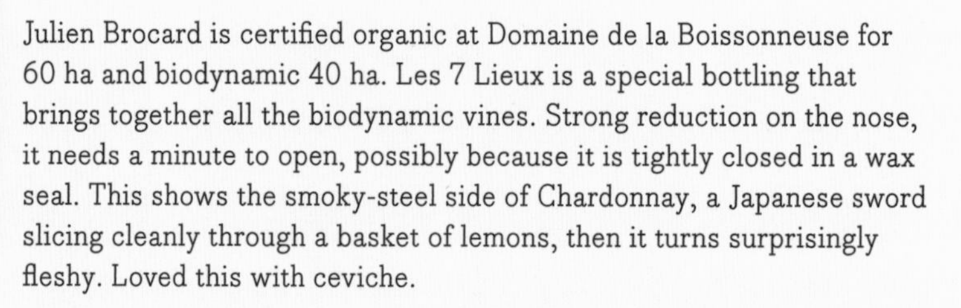

JULIEN BROCARD DOMAINE DE LA BOISSONNEUSE LES 7 LIEUX, CHABLIS 2015, FRANCE

Julien Brocard is certified organic at Domaine de la Boissonneuse for 60 ha and biodynamic 40 ha. Les 7 Lieux is a special bottling that brings together all the biodynamic vines. Strong reduction on the nose, it needs a minute to open, possibly because it is tightly closed in a wax seal. This shows the smoky-steel side of Chardonnay, a Japanese sword slicing cleanly through a basket of lemons, then it turns surprisingly fleshy. Loved this with ceviche.

VINE REVIVAL TERRE DE GNEISS MUSCADET-DE-SÈVRE-ET-MAINE 2015, FRANCE

WWW.VINEREVIVAL.COM

I'll just preface this by saying that the winemaker Christelle Guibert is tasting director at *Decanter* magazine in London, meaning she gets to sample many thousands of wines per year, and is putting her well-worked palate to great use with this subtle and beautiful Muscadet. She works with her sister Corinne and fellow Loire winemaker Vincent Caillé of Domaine Fay d'Homme (also biodynamic, and who worked these vines previously). Together they are slowly expanding a project to rescue abandoned vineyards in the Muscadet region. Vinified in concrete eggs, you'll find elegant flavours: some hay and grass notes, against white flowers. Give it a minute for a second wave of fuller, deeper flavours, but this is delicacy itself – apricot blossom and soft apples, with a pleasing touch of salinity on the finish. Oh, and Gneiss (pronounced 'nice'), refers to the rocky soil. Begs for seafood.

ZUSSLIN GRAND CRU PFINGSTBERG 2013, FRANCE

WWW.ZUSSLIN.COM

The Zusslin family has racked up an impressive 13 generations at the helm of this vineyard, with brother and sister Jean-Paul and Marie running things since 2000. This is a dry (and then some) Riesling, with sharp almost zingy clarity and gorgeous, razor-sharp citrus flavours. Upright, sexy, tight and focused, with ripples of steel and huge persistency. With excellent tension, this pretty much floats above your palate. A great match for a crayfish salad. Demeter-certified.

DR. BÜRKLIN-WOLF WACHENHEIMER RIESLING TROCKEN 2013, GERMANY

WWW.BUERKLIN-WOLF.DE

World-class dry Rieslings. The largely volcanic-soil vineyards are all farmed without herbicides or pesticides by Bettina Bürklin von Gurdaze and her husband Christian. This gives a perfect introduction to the rich yet razor-sharp quality of their Rieslings, rich stone fruits meet gentle white pepper meet a rough slate wall that your palate basically climbs up towards the finish line. Perfect with pan-fried snapper with lime and jalapeño. Biodyvin-certified.

Peter Jakob & Peter Bernhard Kühn

Weingut Peter Jakob Kühn

MÜHLSTRASSE 70, 65375 OESTRICH-WINKEL, GERMANY
WWW.WEINGUTPJKUEHN.DE

We are into 11th and 12th generation winemaker territory at this estate, with Peter Jakob Kühn working alongside his son Peter Bernhard, who earned his winemaking stripes in Burgundy as well as in Alsace with Zind-Humbrecht before coming home. Daughter Sandra Kühn-Quenter also joins them – she and Peter Bernhard graduated from the prestigious Geisenheim University.

The estate was founded in 1786 by Jacobus Kühn, and seems to be forever (and deservedly) winning awards for its brilliant range of wines. They are interested in a 'very true and pure' approach to dry Riesling, as Peter Bernhard told me, and the wines are stunning – the perfect example of how rich and round Riesling can be, while maintaining a steely, almost painfully flinty, tense core. Plus, they like to say that they also have a 'secret love affair with Pinot Noir' – the vineyards are planted with 90% Riesling and 10% Pinot Noir.

Peter Jakob and his wife Angela have been steadily setting out their stall for biodynamic winemaking for years – they were the first ones in the Rheingau to convert and today are Demeter-certified.

'I like wine to taste and smell the way it used to taste and smell, with natural yeasts. Not with one of the nearly three hundred yeast strains that are artificially produced. I'm trying to plant little seeds in the glass to encourage people to think about how wine should naturally taste.
Peter Jakob

WEINGUT PETER JAKOB KÜHN OESTRICH DOOSBERGER RIESLING GG 2014, GERMANY

The first biodynamic producers in the Rheingau produce this absolutely incredible Riesling from 45-year-old vines grown on southwest slopes giving extremely low yields. Flint-edged citrus that pierces right through to your heart, it's no surprise that the residual sugar level of 2.9g/l in this style of Riesling is basically completely undetectable. This is just a stunning wine, it grips on to your palate and keeps on issuing little commands that keep you hooked.

Opposite: Biodynamic treatments are regularly applied at Weingut Peter Jaokb Kühn.

WEINGUT A CHRISTMANN KÖNIGSBACHER IDIG RIESLING GG PFALZ 2010, GERMANY *

WWW.WEINGUT-CHRISTMANN.DE

This wine is now approaching seven years old and you start to see a richer, more golden colour to the Riesling – one of the world's grapes best suited to long ageing. Saline, saffron and burnt-orange rind right from the off. It hits you with flavour then takes a breath, suspending itself in mid-air while you wait, mouth watering, for the next hit. It is both an excellent apéritif wine with a bowl of olives, but is equally stunning with a well-aged Comté. Demeter-certified.

LA MESMA INDI GAVI DOCG 2015, ITALY

WWW.LAMESMA.IT

I first tasted these wines at the Slow Food Wine Bank in Piedmont (heaven on earth if you have the chance to visit), and have since got to know more of the range, and the three Rosina sisters behind it – Paola, Francesca and Anna. The grapes for this particular bottling are certified organic (with the rest of the estate in conversion) and are vinified with natural yeasts (INDI stands for indigenous yeasts and, they say, the indie music that they listen to in the winery). Expect a light and fresh style, with grippy, citrus flavours. A wine to crack open and enjoy with equally fresh, light flavours – grilled vegetables, antipasti, salami, Sicilian olives, *confit de lemon*.

WEINGUT BRUNDLMAYER GRUNER VELTINER SPIEGEL 1 VINCENT KAMPTAL RESERVE 2014, AUSTRIA * * *

WWW.BRUENDLMAYER.AT

This is produced by Willi Bründlmayer's son Vincent, who picked his favourite section of the vineyard to flex his creative muscles and began working it entirely organically in 2011 (the rest of the estate has since followed suit). I challenge anyone to say that mineral isn't a good description for how a wine can taste after drinking this one. The whole structure just hums with tension, strewn with spiced quince, stone fruit and a lick of intense citrus. Stick this next to a lightly grilled John Dory and enjoy.

Nikolaus Saahs
Nikolaihof

NIKOLAIGASSE 3. 3512 MAUTERN, AUSTRIA
WWW.NIKOLAIHOF.AT

Of all the incredible wines in this book, I think Nikolaihof got more sommelier recommendations than any other, which makes it pretty good news for all of us that the Saahs family was stubborn enough to buy Nikolaihof not once but twice since 1894. It was first requisitioned by the Wehrmacht in 1938 (leaving them a small amount of vines), then under Russian occupation until 1956, when Nikolaus' grandfather bought the winery building back.

'There was absolutely no money for any treatments or fertilizers in the vines, so he simply didn't use them. And when there was more money, he already knew that the vineyard didn't need them,' says Nikolaus today. 'Then my parents began working biodynamically by 1971, at a time when most of the neighbours thought they were crazy to do so.

I was born and raised in this environment, making it totally normal for me to follow this philosophy. Today we use no herbicides, artificial fertilizers, pesticides or synthetic sprays and replace them instead with stinging-nettle manure, valerian drops, horsetail tea and their own specially produced Demeter preparations. Our thought is simply to imbue the wine with as much strength and energy as possible, while leaving nature to its own devices as far as possible.'

'In the long run, I see no other alternative to biodynamic farming to maintain the quality of our soils.'
Nikolaus Saahs

NIKOLAIHOF WACHAU VINOTHEK RIESLING 2000, AUSTRIA

A family domain owned by Nikolaus and Christine Saahs, this has the distinction of being the first biodynamic wine estate in Europe. The Vinothek is rich, deep, steely but textured, almost round, fleshy apricot fruit, showing the full flexibility of the Riesling grape with flashes of steel wrapped up in rich lemon. I have chosen an older Riesling to show just how stunningly they age, but the brilliant Gerard Basset suggests a younger Grüner, if you would rather.

'A very complex dry white wine with a lot of energy combined with superb aromas of citrus and stony fruit as well as some touches of celeriac. Magnificent with a very light ceviche of scallops.'
Gerard Basset MW, Best Sommelier in the World 2010 on Nikolaihof Hefeabzug Grüner Veltliner 2013

Johannes Hirsch

Weingut Hirsch

HAUPTSTRASSE 76, 3493 KAMMERN, AUSTRIA
WWW.WEINGUT-HIRSCH.AT

The vineyard here dates back 500 years, with the Hirsch family its caretakers since the 1870s. It is set across three sites that are considered among the best in Austria, and all three have extremely different soils. The skill with which Johannes works these soils into his astonishing range of Rieslings and Grüner Veltliners has made him one of the most sought-after winemakers in Austria.

Quietly spoken, with a gentle but quick sense of humour, he doesn't miss a detail in pursuit of great wine. In 1999 he took the somewhat radical step of pulling up all his red vines to concentrate on only these two white grapes, and was the first producer in Austria to bottle using the Stelvin screwcap closure. I love to contrast that with the traditional methods used in the vineyards. His father, for example, began using a rather special compost over 30 years ago by trading hay from his own fields with a local cheesemaker called Robert Paget. The hay was then used to feed goats and water buffalo that would eventually provide the local region with delicious cheese – and their manure came back to Hirsch for composting the vineyards. It was this process that set the stage for their later conversion to organics and biodynamics.

Johannes started winemaking school (at the world's oldest wine school) at the age of 14 in 1985 – right in the middle of the Austrian wine crisis.

'At that time it wasn't exactly cool to be a winemaker,' he has said, of that time. But in fact it turned out to be exactly the right time to be starting, as there was a renewed focus on quality. Johannes made the most of this, by first choosing to travel and experience winemaking in New Zealand, Australia, California and South Africa.

> 'At first when I came back I wanted to change everything – thought maybe my future was Chardonnay. But at some stage I realized that Grüner and Riesling were the right grapes for our land, so that was what I focused on. Then around fifteen years ago, I started tasting biodynamic wines and found this extra level of salinity, of vitality in them, and knew I wanted that in our wines.'
>
> *Johannes Hirsch*

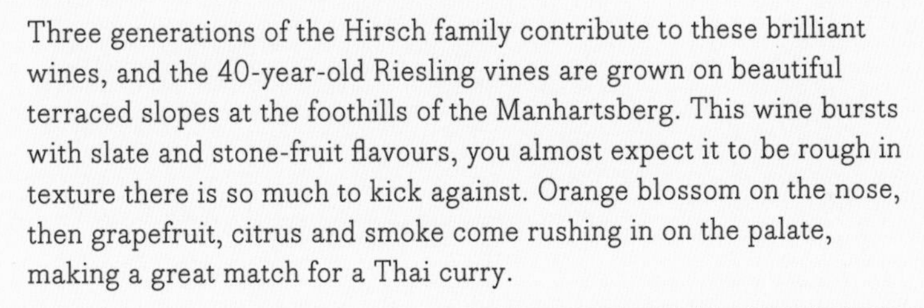

WEINGUT HIRSCH ZÖBINGER GAISBERG KAMPTAL RIESLING 2014, AUSTRIA

Three generations of the Hirsch family contribute to these brilliant wines, and the 40-year-old Riesling vines are grown on beautiful terraced slopes at the foothills of the Manhartsberg. This wine bursts with slate and stone-fruit flavours, you almost expect it to be rough in texture there is so much to kick against. Orange blossom on the nose, then grapefruit, citrus and smoke come rushing in on the palate, making a great match for a Thai curry.

Opposite: The Hirsch vineyard dates back 500 years.

GUERILA ZELEN 2015, SLOVENIA

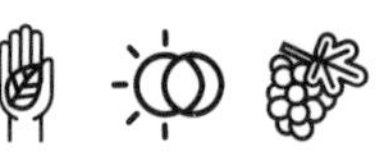

WWW.GUERILA.SI

Winemaker Martin Gruzovin and owner Zmago Petrič concentrate on the rare grapes of Zelen and Pinela – there are just 60 ha in total grown in Slovenia, found in the Vipavska Dolina region ('and nowhere else on this planet,' as Martin puts it). The Zelen is grown biodynamically on mountain slopes in the Vipava Valley, at 450m (1476ft) above sea level, made with a blend of ripe and overripe grapes. How do I describe this wine? The name means green and it has a beautifully bright, clean edge to it, with apricot blossom building slowly over the palate, and a soft herbal edge of rosemary and thyme that would make it an excellent match to a fresh goat's cheese. This is a special and unusual wine that should be more widely known.

DOMAINE SPIROPOULOS MOSCHOFILERO PDO MANTINIA 2013, GREECE

WWW.DOMAINSPIROPOULOS.COM

The Spiropoulos family pretty much single-handedly pioneered organic winemaking in Greece, converting their Peloponnese estate back in 1993. This is a lovely delicate wine from the highly aromatic pink-skinned Moschofilero grape, grown at 650m (2133ft). With only free-run juice, natural yeasts and no secondary fermentation the wine keeps its freshness and focus intact. Prepare to be surprised at the rose-meets-confit-lemon flavours – I'd happily stick with grilled halloumi for this, because why overcomplicate things?

'Floral and elegant aromas of lemon blossom and rose, followed by fresh flavours of citrus, peach and white fruits.'
Jules Aron, The Healthy Bartender, New York

SCLAVOS VINO DI SASSO ROBOLA DE CÉPHALONIE, GREECE

KECHRIONA, PALIKI PENINSULA

From a family of Russian-Greek descent, this estate is certified organic and uses biodynamic practices. Located 850m (2789ft) above those sparkling blue Greek seas on blinding white limestone (the name means Wine of Stone), this isn't hugely aromatic on the first nose, it all comes in on the second wave – then a smoky, grilled edge to the white flower and gentle honeyed notes appears, with an inviting mineral edge. This is a hugely flexible food wine that I enjoyed with a heaped plate of breads, olives and aubergine pâté.

Opposite: Domaine Spiropoulos in Greece dates back to 1870.

TESTALONGA EL BANDITO SWARTLAND 2015, SOUTH AFRICA

@MADEFROMGRAPES

Craig Hawkins is making this wine in his own image – right down to the idea that all the labels are photos taken by friends or a random image that has caught his eye. He doesn't own his vineyards but works with the same growers every year. The 100% Chenin Blanc grapes are from hand-pruned old bush vines, whole-bunch pressed, and left on the lees for ten months without stirring. This has incredible persistency, a beautiful wine that begs you to sit down, open a book and just relax.

GROSSET POLISH HILL CLARE VALLEY RIESLING 2015, AUSTRALIA

WWW.GROSSET.COM.AU

Polish Hill

2015

1981

35YEARS

This wine, from the Polish Hill vineyard at 460m (1510ft) altitude, is frankly wonderful. Distinct, tight, steely, with a pointed finish not overdone, it has impressive hold from beginning to end. White pear flesh, slightly peppery, touches of melon. Would be a good with simple but slightly spiced flavours like roasted sea bass and ginger.

PYRAMID VALLEY VINEYARDS LION'S TOOTH CHARDONNAY 2014, NEW ZEALAND

WWW.PYRAMIDVALLEY.CO.NZ

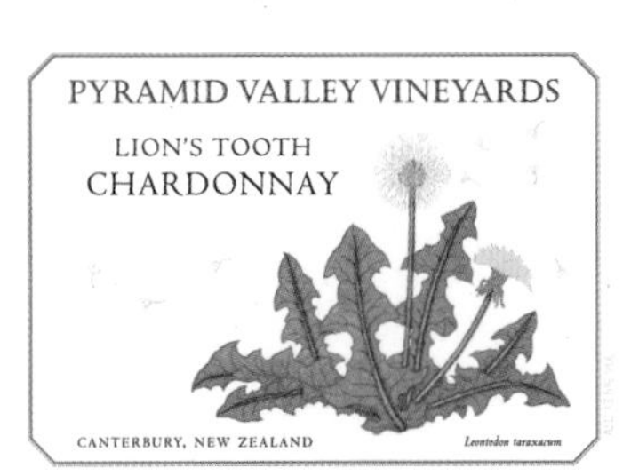

This was the first vineyard in New Zealand (and one of the few globally) to be biodynamic from the very first plantings. It was founded by Mike and Claudia Weersing in 2000, after 15 years of looking for the right site. Named after the French Dent de Lion dandelion, it refers to a plant, which is particularly useful in biodynamics as part of the 506 preparation. Unfiltered, unfined, with tiny amounts of sulphur, this is spicy, succulent, poised and extremely complex – sweet grilled nectarines against a herbal, floral backdrop. It really is a surprising wine that has a smoky core and lovely minerality, and I think a great pairing for grilled white fish or a fresh crab salad.

'Would go perfectly with Salmon poké bowl with crispy shallots, chili mayonnaise.'
Caleb Ganzer, Wine Director and General Manager, La Compagnie des Vins Surnaturels, New York

Opposite: Pyramid Valley practiced biodynamic farming from its inception.

Wine Cocktails

by Jules Aron, The Healthy Bartender
www.thehealthybartender.com

Almost any flavour can be highlighted by wine. The array of botanicals makes the complexity of flavours practically limitless. Artisan wines work particularly well in cocktails because of their cleaner, crisper flavours.
NOTE: a standard small serving of wine is 125ml (4fl oz).

Tempranillo Old Fashioned

90ml (6 tbsp) Tempranillo
30ml (2 tbsp) black cherry juice
15ml (1 tbsp) lemon juice
2 dashes cherry bitters
dash of bourbon
ice
2 bourbon soaked cherries, for garnish
1 orange twist, for garnish

To make the bourbon soaked cherries:
340g (12oz) black cherries
120ml (4fl oz) bourbon
orange peel from 1 orange
120g (4½oz) maple syrup
100g (3½oz) orange juice
a drop of vanilla extract

Make the bourbon soaked cherries in advance: heat all the ingredients, except the cherries, together for 5 minutes, then pour over the cherries and leave in the fridge to cool for 3 days.

In a mixing glass, combine the Tempranillo, black cherry and lemon juice and bitters. Before pouring, add a small dash of bourbon to a chilled red wine glass, gently swirl the liquid to fully coat, then pour excess out. After this rinse, now fill your mixing glass with ice, stir well, and strain into the red wine glass.

Garnish with bourbon cherries and an orange twist: thinly slice a piece of orange peel (no pith) and twist to introduce some of the peel's oils into the drink.

Fresh Orchard

90ml (6 tbsp) chilled Chardonnay
45ml (3 tbsp) fresh Granny Smith apple juice
15ml (1 tbsp) Calvados
15ml (1 tbsp) lemon juice
ice
1 lemon twist and 1 thyme sprig, for garnish

In a mixing glass, combine the Chardonnay, apple juice, Calvados and lemon juice. Fill the glass with ice, stir well and strain into a chilled white wine glass. Garnish with a lemon twist and a thyme sprig.

Everything Coming Up Roses

1 x 750ml (1¼ pint) bottle chilled dry rosé
240ml (9fl oz) chilled brewed hibiscus tea
120ml (4fl oz) fresh lemon juice
30ml (2 tbsp) rose water
120ml (4fl oz) simple sugar syrup infused with honey and passion fruit
ice
230ml (8fl oz) sparkling water
strawberries, orange and lemon wheels, for garnish

In a punch bowl, combine the rosé, tea, lemon juice, rose water and syrup. Fill the punch with ice and stir well. Stir in the sparkling water and serve in chilled wine glasses.

Garnish with the strawberries, orange and lemon wheels: slice the strawberry down the middle but not into two and place onto the side of the glass, then slice a thin round of orange and lemon and again cut in to it almost to the middle point so that it sits on the side of the glass.

02

Rich & Round Whites

Franck Moreau Master Sommelier (MS)

Group Sommelier at Merivale, Sydney, Australia

I was born in France, and grew up in a small village in Beaujolais, surrounded by vineyards. These childhood memories may be partly why I love biodynamic and organic wines so much – they are the result of intense focus and care that the wine-growers give to their vineyards, and show the passion needed to make a truly terroir-driven wine.

I have now been living in Sydney since 2004, working across a number of restaurants where we celebrate many different styles of wine and food. These richer whites are always a wonderful category for food-lovers to get excited about, as they are so diverse and come from a large variety of grapes and styles.

One of the key things to bear in mind with richer-style whites is their texture. We have the tendency to describe all types of fruits and other characters in wines but forget about the structure and mouthfeel – both of which are crucial to our enjoyment of them, and in picking successful food matches.

The range of grape varieties may vary from the classic Chardonnay, Grüner Veltliner, Chenin Blanc, Albariño, Viognier, Sémillon, Marsanne, Roussanne, Grenache, to the more obscure varieties such as Caricante, Cattaratto, Nosiola, Godello, Albillo, Malagousia, to name a few.

Why do these wines match so well with food? Here are some of my tips:

textural and richer styles work well with simple rock, fleshy and fatty fish;

fuller-bodied wine needs food accompanied by creamy and strong-flavoured sauces and also matches well with poultry, goose and lighter meat dishes;

medium-bodied and aromatic grapes, such as Gewürztraminer or Viognier, with lower acidity, work with light seafood with an Asian influence, such as grilled scallops with a ginger sauce;

skin-contact wines, such as orange wines, express strong 'umami' flavours and work well with textural dishes, such as Hawaiin raw tuna salad, simple cured meat, or even game and mushroom consommé;

wines with high acidity, such as Chenin Blanc, need rich/oily food like salmon, chicken curry or classic fish and chips;

white wine with cheese is great – mostly goat's cheese and washed-rind cheese, such as Pont l'Évêque, Époisse de Bourgogne or Chabichou du Poitou.

And here are a few simple rules to always consider when matching food and wine:

match the intensity of the food with the weight of the wine;

never overpower the dish with the flavours of the wine;

the acidity in the wine needs to cut through fatty dishes;

spicy food needs wine with a bit of sweetness;

the wines should bring an extra extravaganza to the dish or some *je ne sais quoi*!

contrast can be interesting but is also difficult to get right.

Franck's Wine Picks

COS PITHOS BIANCO, SICILY

DAMIEN LAUREAU SAVENNIÈRES LES GENÊTS, LOIRE, FRANCE

NGERINGA CHARDONNAY, ADELAIDE HILLS, SOUTH AUSTRALIA

PHEASANT'S TEARS CHINURI, GEORGIA

TERROIR AL LIMIT TERRA DE CUQUES, PRIORAT, SPAIN

John Williams
Frog's Leap

8815 CONN CREEK RD. RUTHERFORD, CA 94573, USA
WWW.FROGSLEAP.COM

John Williams grew up on a small-scale dairy farm, coming to Napa in 1975 after studying oenology at UC Davis, 'back in the days when every single grape was dry-farmed and no one was using shortcuts in the vineyards'.

It is an approach that he has stuck to ever since, even at times when he felt like the only person in the valley to be doing so. My guess is it helped that John is not somebody who is afraid to say what he thinks.

'What astounds me is that anyone would farm any other way than organically,' he says bluntly. 'It seems counter intuitive to use synthetic fertilizers or chemical pesticides that will affect the connection between a vine and its soils, when your whole business basically depends on that connection. It's like a famous chef not caring about his ingredients – it just seems a no-brainer.'

His early career more than proved the worth of this approach. He apprenticed at Stag's Leap at a time when it won first place for the red wines in the 1976 Judgement of Paris tasting, which set California wines on to the path to international renown. He says that his experience with Warren Winiarski as mentor convinced him of the 'fundamentals of great wine being focused on the vineyard first'.

He says he is drawn to wines with simplicity ('because that is the hardest thing of all to achieve') and a sense of place.

> 'It was the great terroir of Napa that first persuaded me to stay here, and I always try to keep the concept of its voice in my mind when making wine.'
> *John Williams*

FROG'S LEAP SAUVIGNON BLANC RUTHERFORD NAPA VALLEY 2015, USA

Frog's Leap wines stand out for their integrity and restraint. This 100% Sauvignon Blanc has a beautifully fresh 3.22pH, and is full of fragrant citrus, gentle white peach and lemongrass. Unoaked but with a sense of grip as the flavours deepen over the palate, it would match perfectly with a wild mushroom tart. The wines have been certified organic since 1989, the process started in 1981 on opening, and they have never irrigated. The winery building is LEED certified for sustainability.

Opposite: The rural charms of Frog's Leap winery in Napa.

Ivo Jeramaz
Grgich Hills Estate

PO BOX 450, 1829 ST HELENA HWY, RUTHERFORD,
CA 94573, USA
WWW.GRGICH.COM

Grgich Hills, in the eyes of almost all of us, belongs to Miljenko (okay, Mike) Grgich. But today it is his nephew Ivo who runs the property as both general manager and winemaker. And although Mike has been living in the US since 1954, Ivo grew up in Croatia, like his uncle before him, and both have been deeply influenced by the traditional methods of farming that surrounded them as children.

'When my uncle was young his family produced all their own food, including wine. Farming wasn't called organic, biodynamic, fish-friendly or whatever. It was just farming, and that was how he learned. I learned from him, and strongly believe, as most European winemakers do, that great wine is made in the vineyard. The flavours are in the grapes, so why would we ruin that with aromatic yeasts or too much new oak? The job of a winemaker is not to create wine but to preserve the authenticity of those flavours from the vineyard.

'Mike taught me how to use common sense. Growing grapes is like raising kids – and I have six of them, so I know something about it! It's about observing everyday, and knowing that what is right for one is not always right for the other. There are people who hire consultants, who put probes in the vineyards and monitor from their computers. Good luck to them... but it's not for me.'
Ivo Jeramaz

He has also learned something along the way about family businesses. 'When I first arrived, I used to hate that people knew I was related to Mike because of my accent. But now I understand the strength of family in the wine business. My daughter Maria has just joined us, and she will be the third generation Grgich working the vineyards of California.'

GRGICH HILLS ESTATE PARIS TASTING COMMEMORATIVE CHARDONNAY RUTHERFORD 2013, USA

The label refers to the legend that is Mike Grigch, the man who grew up in Croatia (then Yugoslavia) and became winemaker at Chateau Montelena in Napa when the estate won the 1976 Judgement of Paris white tasting, which basically reset the rules for American winemaking. This is Meursault-style Chardonnay, rich and creamy with some oak toasting, and yet with a steely core. Lemon, apricot, toasted pine nuts, with a touch of sweet oak that will age and is of excellent quality. It can stand up to rich foods, from fish pie to a bouillabaisse or creamy mushroom risotto. Certified Stellar Organic.

Opposite: Grgich Hills is located in Rutherford in the Napa Valley.

Pedro Marques
Vale Da Capucha

LARGO ENG. ANTÓNIO BATALHA REIS,
2565-781, PORTUGAL
WWW.VALEDACAPUCHA.COM

There have been four generations of winemakers in his family before Pedro Marques, but he was the first one to have studied winemaking, and the first to really concentrate on shaking up the traditions of the area. He joined his father Alfonso in 2009, and switched the focus to white wines, which he believed were better suited to the vineyard's location near to the Atlantic Ocean (they are 8 km/ 5 miles away along the Lisbon coast). He also made the switch to organics, becoming certified in 2012, to better express the terroir.

These are brilliant wines, brought alive by Pedro's disarmingly honest approach. He admits that he has made plenty of mistakes along the way, experimenting with plantings of Viognier, for example, and losing serious parts of the harvest while the vineyard was adapting to organics. But he is pretty stubborn when he needs to be, and the wines are now proving that it was worth it.

VALE DA CAPUCHA ARINTO DOC TORRES VEDRAS 2015, PORTUGAL

This is single-vineyard wine from Vale da Capucha, made from 100% indigenous grape Arinto. Old barrels give minimum oak impact, and the wine is fermented with natural yeasts and kept on the lees to build body and texture. Thought-provoking and unusual, this wine focuses more on wild herb flavours than easy fruitfulness. Give it a minute because it seriously improves in the glass; the energy levels rise, the sappy saline grip becomes more evident, eventually beautiful citrus and pineapple appear against the flowers and herbs; it turns out to be totally moreish.

Opposite: Vale da Capucha is located only a few miles from the Atlantic coast in Portugal.

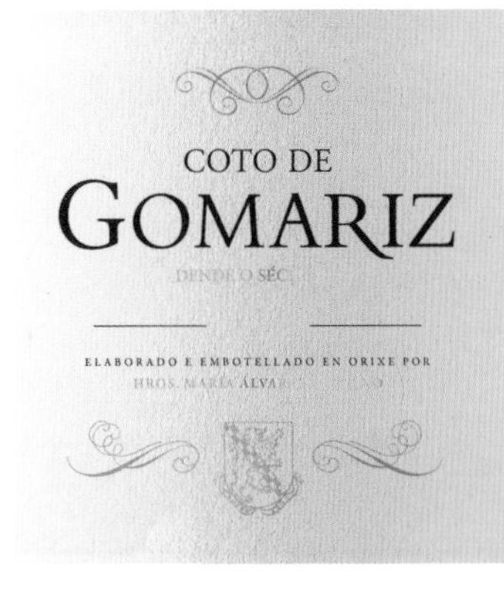

COTO DE GOMARIZ BIANCO RIBEIRO 2015, SPAIN

WWW.COTODEGOMARIZ.COM

A lovely and unusual blend of Treixadura, Godello, Loureira and Albariño – four grapes native to the Galicia region. The vineyard sits at around 250m (820ft) altitude, and benefits from the natural low yields that come from its granite, schist and clay soils. The wine is vinified in stainless steel at low temperatures, has no oak and is bottled on the biodynamic calendar's flower days. Winemaker Xosé Luis Sebio manages to pack an absolute punch of flavour: grapefruit and lemon zest abound on the palate, with a salty-slatey twist on the finish, in a kind of wine version of a gin and tonic. A brilliant match to a bulgur wheat or quinoa salad, where it can play off the texture of the food.

LAUREANO SERRES MONTAGUT MENDALL FINCA ABEURADOR BLANC TERRA ALTA 2015, SPAIN

WWW.SERRES.NET

When you're called 'hardcore' by natural wine expert Alice Feiring, you know you are doing something right. Laureano Serres lives in the Terra Alta region of Catalonia (apparently where Pablo Picasso spent his summers) and has a reputation of being a mad genius. This Abeurador is made with 100% Macabeo, fermented using wild yeasts in an amphora with no temperature control or added sulphur. It has an earthy, slightly oxidized character to the almond, melon and white flower notes, but just gets better and better in the bottle and glass; my bottle was open for five days and it was still fresh and full of energy.

PALACIO DE CANEDO GODELLO BIERZO 2015, SPAIN

WWW.PRADAATOPE.ES/EN/THE-WINES

If you get a chance to visit this place, take it – restaurant, hotel, vineyard, winery, a general fantasy spot in northwestern Spain that begs for organic viticulture and gets it in spades. The Godello grape had almost died out in this part of Spain a few decades ago, but a small group of growers – including Prada here – nurtured it back to health. This is stony, herbal, intense rosemary over grilled lemons, touches of white-pepper spice with tropical fruits, smoky and delicious.

'Just love this Godello. We have a poached lobster ceviche that is fab with it. Leche de tigre sauce (so lime juice, onions, chile, salt, pepper and lobster stock) of course. But it's the sweet corn kernels that we add in that really makes it work.'
Jeff Harding, Sommelier, The Waverly Inn, New York

QUINTA DA MURADELLA GORVIA BLANCO MONTERREI 2011, SPAIN ✱ ✱

LUIS ESPADA, 99, 32600 VERIN, MONTERREI

This might be a cult wine, but it is also the real deal – the product of passionate and talented winemaker José Luis Mateo and oenologist Raul Perez doing their thing. This Gorvia is rich without being heavy; you feel it roll around in your mouth, with peach blossom, citrus and nectarine flavours, a touch of bitter almond, salty freshness and ginger spice on the finish. Made from the Doña Blanca grape grown on schist/slate soils, it is fermented with natural yeasts in 1500-litre oak vats, one-third fermented on the skins and aged in barrel for eight months. A rich shellfish dish like baked crab would stand up well.

TERROIR AL LIMIT TERRA DE CUQUES PRIORAT 2014, SPAIN ✱ ✱ ✱

WWW.TERROIR-AL-LIMIT.COM

Made by Dominik Huber and Eben Sadie, the name translates as 'Bug's Earth' in honour of the fireflies that light up the landscape of this area at night. It's an unusual blend of 80% Pedro Ximénez and 20% Moscatel de Alejandria macerated and fermented on the skins using whole clusters. This is spine-tingling wine, racy, zesty, roasted almonds meets lemon curd and lime zest. Wow.

'I love this wine with snapper fillet with ginger and shallot sauce.'
Franck Moreau MS, Group Sommelier, Miravale, Sydney

Above: The hotel and restaurant at Palacio de Canedo in Spain.

Telmo Rodriguez
Remelluri

LABASTIDA WINERY & OFFICES, CARRETERA RIVAS DE TERESO, S/N, 01330 LABASTIDA, ALAVA, SPAIN
WWW.REMELLURI.COM/EN

One of my favourite memories in 15 years of writing about wine is sitting on a stone bench on the Remelluri estate, glass of red in hand, plate of estate-grown olives to one side, watching the sun set over vines and the mountains behind, the whole place heavy with the scent of wild anise, thyme, rosemary and sage.

Telmo Rodriguez is sometimes called Rioja's Prodigal Son, because he left his family property behind for a good ten years while making his own wines around Spain (he still does, under his own name, and if you see them on a wine list, don't hesitate). He came home wanting to 'reflect the real Remelluri, and the original taste of Rioja,' as he puts it.

Today he works with his sister Amaia, and has, as he somewhat ruefully admits, 'increased the costs of making Remelluri drastically'. This means he no longer buys in any grapes from outside growers and concentrates instead of ensuring the wine is a reflection of Rioja's history. One plot in the Remelluri vineyard is planted with over 20 historical varieties, used as a nursery to maintain genetic diversity. Bush training is being reintroduced, as are terraces and a variety of concrete tanks; oak barrels and large-sized oak vats – some new, some old – are used to age the wines.

Telmo – who, let's face it, adds a touch of realism to the term 'poster boy' – is now taking his beliefs further, forming a group of terroir-driven winemakers across Spain, encouraging others to connect with the indigenous grapes of their area, and to put the emphasis on the diverse terroirs of the country.

REMELLURI BLANCO RIOJA 2014, SPAIN * * *

This is a field mix of Viognier, Chardonnay, Roussanne, Marsanne, Sauvignon Blanc, Garnacha Blanc and Moscatel del Pais, even winemaker Telmo Rodriguez doesn't know the exact percentages. Taste-wise, it's equally difficult to pin down. There's a rich squeeze of citrus and fleshy apricot juice, a clutch of the many wild anise and sage herbs that grow on the mountainside behind the winery, all wrapped up with a searing minerality. This is a wine I buy whenever I can, and fail miserably to hold off opening. Extremely flexible with food, one of my favourite matches: grilled squid and pepper salad.

Above: Remelluri estate is based along the slopes of the Sierra de Tolono mountains in Rioja.

Alexandre Bain
Domaine Alexandre Bain

BOIS FLEURY, 18 RUE DES LEVÉE,
58150 TRACY-SUR-LOIRE, FRANCE
WWW.DOMAINE-ALEXANDRE-BAIN.COM

So, I have two stories about Alexandre Bain. The first dates back to 2015, when I was writing a news piece about him being stripped of his right to put his appellation on the label because he felt – although that was not the reason they gave – that the local authorities believed his low-sulphur, minimal-intervention wines no longer tasted typical of the appellation. (He has since won on appeal.) I followed up our conversation by buying a couple of bottles to try for myself, and loved them.

And then a few months ago, when researching this book, I asked a sommelier in the brilliant Vins Surnaturels wine bar in Paris to name one winemaker in France whose natural wines you could totally, completely trust to not disappoint and he said, without hesitation, Alexandre Bain. And he's not the only one – Alexandre features on the wine lists of plenty of restaurants who truly care about sourcing excellent raw materials, from Noma in Copenhagen and L'Arpège in Paris.

He founded the winery in 2007, becoming a winemaker because he loved it not because he was taking over a family property (although his grandfather did once have vines in the region), and converted it to organics immediately.

> 'I harvest late, usually around two weeks later than my neighbours, and my yields can be around half the size. This means the flavours are more honeyed, less mineral than some classic Pouilly-Fumé. But I do everything I can to ensure they are a true reflection of the land that grows them.'
> *Alexandre Bain*

ALEXANDRE BAIN CUVÉE PIERRE PRECIEUSES POUILLY-FUMÉ 2016, FRANCE

This wine is 100% Sauvignon Blanc, picked as late as winemaker Alexandre Bain dares, fermented with natural yeasts, just shining with juicy fresh peaches and grapefruit. This would be the perfect partner to any number of simple fresh ingredients, from a warm lentil salad to a plate of freshly sliced salmon sashimi.

BENEDICT & STÉPHANE TISSOT LES GRAVIERS ARBOIS 2015, FRANCE * * *

WWW.STEPHANE-TISSOT.COM/EN

I fell head over heels in love with this Chardonnay the first time I tried it. Stéphane Tissot is one of the pioneers of biodynamic winemaking in the Jura and Les Graviers is an amazing example of what this region can produce – and it's not even the top priced of his range. Laughably easy to gulp down, so rich in flavour and yet mineral, clean, pure. It contracts on the finish, thrillingly tense and keeps you held there for a good five minutes. Candied lemon is cut through with a twist of concentrated lime and cut slate. I drank this with sumac chicken and am salivating at the memory; it's a wine that makes you want to call your friends to share the good news.

CHÂTEAU DE PIBARNON NUANCÉS ROSÉ BANDOL 2015, FRANCE

WWW.PIBARNON.COM

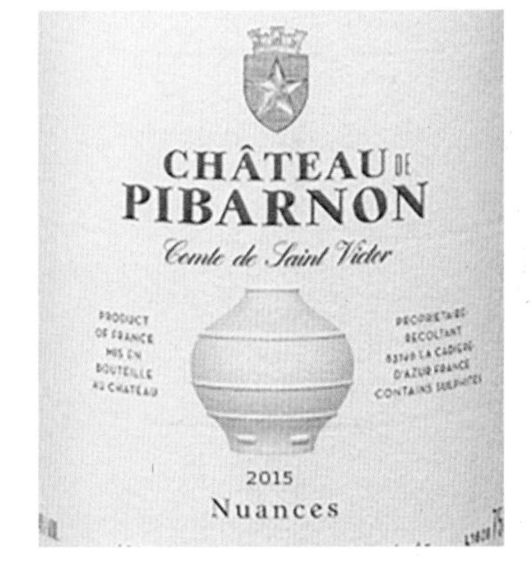

I get a shiver of excitement just thinking about turning into the narrow pathway that takes you down to this estate. Bandol is known for its *garrigue*-studded red wines, but it is also a place where Provence rosé gets serious. Eric de Saint Victor's 100% Mouvèdre shows you how, vinified with natural yeasts in a mix of 3000-litre Stockinger Austrian wooden barrels and sandstone jars. Most Provence rosés are out the door by Christmas following harvest, but this one hangs on until the following summer, developing complex and refined flavours of crushed roses, rich berry fruits and gentle spices. A sundowner, perfect for blackened cod or Cajun-spiced chicken.

CHÂTEAU ROMANIN ROSÉ BEAUX DE PROVENCE 2015, FRANCE

WWW.CHATEAUROMANIN.COM

This balances real tension with the elegant red-berry fruits that you look for when cracking open a pink wine from Provence. From a blend of Grenache, Counoise and Syrah, there is clear salinity on the finish, with elderflower and redcurrant sorbet adding to the mouthwatering effect. Owned by the Charmoule family since 2006 (who moved here after selling Château Montrose in Bordeaux's St-Estèphe), this is an idyllic spot, set in the Alpilles Natural Park. Olive oil and honey are also made. Perfect with the salty tang of tomatoes, feta and olives.

Overleaf: The buimdings at Château Romanin in Provence work in harmony with the environment.

Nicolas Joly
Coulée de Serrant

CHÂTEAU DE LA ROCHE AUX MOINES,
49170 SAVENNIÈRES, FRANCE
WWW.COULEE-DE-SERRANT.COM

'Some wines are just too straight. Drinking them is like driving on a motorway.'

This is classic Nicolas Joly. A philosopher, an intellectual, a farmer, an ex-investment banker with an MBA from Columbia turned biodynamic evangelist.

Nicolas returned to his family estate in the Loire in 1977 and at first brought all his experience of the wider world with him. 'That means I thought I knew best,' he says, 'and listened to advice that weedkillers would save us a ton of money each year.' He watched with dismay as his vines, and the biodiversity within them, seem to die before his eyes, before picking up a book, more out of curiosity than conviction, which led him to biodynamics. Walk through his vines today, and you're constantly accompanied by a pack of sheep, or a few curious goats, or Maruis the shire horse. There are even chickens around here, although I didn't see them on my last visit.

'Animals are part of the essential balance of the vineyard, the vines draw energy from having them near.'
Nicolas Joly

An afternoon with Joly is a constant battle between wanting to ask questions and wanting to listen. As with many true believers, he can be scathing of those whose attempts don't meet up to his standards (zero sulphur in winemaking, for example, enrages him if it is replaced with something more harmful such as an extremely heavy filtration), and yet he travels widely to help young winemakers. In 2001 he was one of the founders of the Renaissance des Appellations, assuring quality, authentically made wines, with Anne Claude Leflaive, Pierre Morey, Olivier Humbrecht, Philippe du Roy de Blicquy (Domaine de Villeneuve).

COULÉE DE SERRANT CLOS DE LA COULÉE DE SERRANT 2014, FRANCE * * *

La Coulée de Serrant is one of three wines that Nicolas Joly makes on his property Château de la Roche aux Moines. This is a stunning, heart-stopping wine made with the Chenin Blanc grape, combining *fleur d'orange* with white truffle, saffron, candied lemon, gunsmoke and gentle white flowers. Wild yeast, natural fermentation, extremely low sulphur: a natural way before the term natural wine came about. Try with mullet and a saffron-laced sauce. Demeter and Biodyvin-certified.

Opposite: Autumn in the vineyards is a beautifully colourful time.

Jean-Charles Abbatucci
Domaine Comte Abbatucci

MUSULELLO, 20140 CASALABRIVA, FRANCE
WWW.DOMAINE-ABBATUCCI.COM

It took ten years, Jean-Charles Abbatucci admits, for the effects of converting to biodynamics to truly take effect, 'but from the very first year I began to see the differences. It is essential to be closer to the plants once you start farming this way, and you are learning every single year simply by observing. The journey is exterior, as you watch the vines, but also internal with yourself.'

I defy you to spend time with Jean-Charles and not want, just a little bit, to give up your day job and buy a piece of wild land to convert to biodynamic farming. I was practically packing my bags by the end of a half an hour phone conversation, never mind a few days spent in his Corsican paradise.

> 'Nature does best when man doesn't get involved.'
> *Jean-Charles Abbatucci*

Today all of the buildings on his estate are built from wood, and everything is self-sustaining. He has a collection of old-vine stock gathered from around the island in the 60s, and is slowly but surely grafting them onto any Carignan, Cinsault and Grenache vines still in his vineyard. Oh, and if you want to know how deeply this goes for him? Well, you've probably heard of people playing music to their vines in the hope of helping them grow. Jean-Charles does this, but instead of playing recorded music, he brings traditional Corsican singers to serenade his slopes. 'It seems essential to me for the vibration to be the real deal,' he says simply, adding that these are 'just tests'.

But one thing Jean-Charles can't fight? The politics of French wine. The most passionate defender of Corsican terroir and grape varieties (his father started a vine conservatory on the property that Jean-Charles maintains today), he was president of Ajaccio AC for a long time, but left in 2008 because the local rules allowed only 10% of indigenous grapes in the wine. 'It was just too restrictive, too limiting.' So these wines are now Vins de France.

DOMAINE COMTE ABBATUCCI CUVÉE COLLECTION GÉNÉRAL DE LA RÉVOLUTION 2014 VIN DE FRANCE CORSICA, FRANCE ✱ ✱ ✱

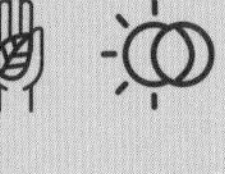

Long before I ever tasted his wines, I kind of fell in love with the wonderfully crazy-sounding Jean-Charles Abbatucci, with his passionate defence of Corsican grapes. A blend of Biancone, Carcajolo Bianco, Paga Debbiti, Riminese, Rossola Brandicana and Vermeninu; extremely pale-straw in colour; rich and creamy with a beautiful seam of acidity and herb-scented citrus, leaving the palate as clean as a whistle. Moreish with red pepper and goat's cheese frittata. Certified Demeter.

DAMIEN LAUREAU SAVENNIÈRES LES GENÊTS 2014, FRANCE

WWW.DAMIEN-LAUREAU.FR

This producer is one of the new superstars of the Loire Valley, with wines made from a vineyard located behind Nicolas Joly's La Roche aux Moines on schist and sand soils. It shows the purity and texture of the Chenin Blanc and is aged for 18 months in barrels. Rich, round but incredibly well balanced with the purity of pear and quince fruit that Chenin can deliver so well. Certified organic since 2012.

'The purity of this wine goes well with shellfish or seafood, my favourite is a blue swimmer crab with Thai flavour garnish.'
Franck Moreau MS, Group Sommelier, Miravale, Sydney

DOMAINE DES COMTES LAFON MEURSAULT-CHARMES PREMIER CRU 2012, FRANCE * * *

WWW.COMTES-LAFON.FR

Described as the world's best producer of white wine, Dominique Lafon is also one of the most influential biodynamic winemakers in Burgundy. In charge since the mid-80s after working in Oregon and California, it was under his direction that this estate turned to organics in 1992, then biodynamics in 1995. This is powerful, concentrated, driven, barely out of the starting blocks – Chardonnay at its very height. The citrus and slate fruit is measured out sparingly, drip by drip, dancing across your tongue as it goes. Pair with morel mushroom risotto.

Above: The stunning colours of Domaine Abbatucci.

Christine Vernay

Domaine Georges Vernay

1 ROUTE NATIONALE 86, 69420 CONDRIEU, FRANCE
WWW.DOMAINE-GEORGES-VERNAY.FR

Christine Vernay spent years teaching Italian at the École National d'Administration in Paris, one of France's famed graduate schools that churns out future politicians, prime ministers and presidents. She would spend her days in lecture halls and her nights listening to recitals from her pianist husband Paul Amsellem, until the two of them swapped Paris for the granite slopes of Condrieu in 1996.

The call back home meant a return to walking, as she says, 'the same slopes and rocky paths as my father and grandfather before me.'

Her father Georges was already known as the Pope of Condrieu for his work in rescuing what had become an increasingly forgotten appellation in the Northern Rhone, dropping down to just 6 ha in the 50s. Even before that, her grandfather Francis planted the first vines of Viognier on these sunbaked hillsides. Without these two generations of Vernay men, one of France's most celebrated white wine appellations – that shows just what the Viognier grape can achieve at the height of its powers – might have been lost.

> 'My father planted on these hillsides at exactly the moment that everyone else was abandoning them and planting fruit trees instead.'
> *Christine Vernay*

It was probably expected that one of Christine's two brothers was going to provide the next generation of winemaker (this is rural France, after all), but they were both busy with their own careers, and faced with the possibility of it being sold, Christine says 'I could feel my roots pulling me home.'

She says, that she learned not only through her father but by 'reading, lots of it', and slowly began to shape her own vision for the property – at first in the cellars and then the vines. She stopped using weedkillers in 2010 (not a decision to take lightly on these slopes, where ploughing by hand is back-breaking and where it is too steep for horses it's right back to a man and a hoe). Today the estate is entirely organic, and is about to begin the process of certification.

DOMAINE GEORGES VERNAY LES CHAILLÉES DE L'ENFER CONDRIEU 2014, FRANCE

The flattering, easy apricot charms of Viognier transform under great producers in Condrieu. All of a sudden the grape becomes unyielding, steely – giving the best Chardonnay a run for its money. This is cold satin in texture, but just when you're relaxing into it, it lifts you up and goes to work with white-pepper spice and razor-blade citrus. The 'Enfer', or hell, comes from vineyard workers who call this particular spot hell for its slopes that make for a tough afternoon's work under the burning Rhône sunshine. Stands up beautifully to slow-cooked lamb.

Jean-François Ganevat
Ganevat

LA COMBE, 39190 ROTALIER, FRANCE

Organic since 1990, biodynamic since 2004, certified with Demeter since 2011, Jean-François Ganevat is about as individual a winemaker as you can hope to find. He cheerfully admits this, saying 'I make lots of wines that don't get approved for not being within the traditional French AC system, but that doesn't bother me. I never want to make wines that follow a recipe, I have no interest whatsoever in that.'

Instead he says, 'We began working in organics by passion, love of the earth and our soils, and when we realized that biodynamics would add a further resonance to the wines, it was the obvious way to go. When you see the impact it makes to the plants, the changes it has in the wine, it really does change the way you think about everything. It also fits in with the French approach to gastronomy – the idea of ensuring the freshest ingredients.'

'But I am not in agreement with every aspect of biodynamics, and I'm not in a cult. We enjoy ourselves, we drink, we have fun, we make natural wines and live our lives. The greatest pleasure that it gives me is the sense of being free, of making wines the way I want to.'
Jean-François Ganevat

DOMAINE JEAN-FRANÇOIS GANEVAT LES CHALASSES VIEILLES VIGNES ARBOIS 2014, FRANCE

This is a richly structured Chardonnay that will grab hold of you and then slowly, deliberately deal out its citrus, grilled herbs and fleshy apriot flavours. Low-yielding old vines, fermented with natural yeasts are then given a long slow ageing of 22 months in large 500-litre oak casks. Jean-François' family have been winemakers since 1650, but used to supplement their income with a herd of dairy cows that provided milk for the local cheese Comté. They must have been on to something, because the two make for a perfect pairing – even better if you add a slice of quince jelly.

Anne-Claude Leflaive
Domaine Leflaive

PLACE PASQUIER DE LA FONTAINE,
21190 PULIGNY-MONTRACHET, FRANCE

WWW.LEFLAIVE.FR

One of true pioneers of biodynamics in Burgundy and across France, I only met Anne-Claude Leflaive a few times before her untimely death in 2015 aged just 59, but she left an indelible impression – on me, on the wine industry, and on the wider world of biodynamic farming.

After growing up mainly in Paris, but spending weekends and holidays at the family vineyard, Anne-Claude moved permanently to Burgundy in 1990 and was the third generation winemaker at Domaine Leflaive – part of a family that had been making wine in the area since the 1700s. She oversaw the estate's conversion to biodynamics in the mid-90s, stubbornly sticking to her beliefs against plenty of opposition from those around her, and became a vocal advocate for the cause, starting among other things the non-profit École du Vin et des Terroirs in Puligny-Montrachet.

This wine school – Anne-Claude was one of a small group of winemaking friends that was behind its creation – offers a way for students to explore wine from an environmental and cultural point of view. She also campaigned against the introduction of genetically modified vines in France, and found the time when not working to sing in a jazz band called Healing the Soul.

Leflaive and her husband Christian Jacques also owned the 10-ha Loire estate Clau de Nell, which continues to be run biodynamically by long-time winemaker, manager and friend Sylvain Potin. It stands as part of Anne-Claude's legacy along with the iconic family property in Burgundy, which is now run by her nephew Brice de la Morandière. The fourth generation of the family in Puligny-Montrachet, he spent many years heading up multinational companies in France and overseas. Brice has pledged to continue his aunt's respect for 'great terroir and humility towards the forces of nature'.

The Leflaive estate makes wines from *grand cru* vineyards including Bätard-Montrachet, Chevalier-Montrachet and Le Montrachet (some of the most prestigious and expensive names in the wine world), but also more reasonably priced areas, such as the Maconnais and Pouilly-Fuissé in southern Burgundy, which Anne-Claude had expanded into – driven, she said, by a desire to provide wines made with biodynamic methods that were more accessible in price, so more able to reach a wider number of wine-lovers.

Opposite: Domaine Leflaive has been part of the story of biodynamics in Burgundy for many years.

DOMAINE LEFLAIVE CHEVALIER-MONTRACHET GRAND CRU 2013, FRANCE * * *

Domaine Leflaive will be forever associated with Anne-Claude Leflaive, the woman who did so much for biodynamics in Burgundy. Honestly, it's been tough deciding which Burgundy estates to put in and which to leave out of this book, because there are just so many brilliant producers working as closely to their terroir as possible. But there was never any doubt that this name was going to appear. I almost went with the Puligny-Montrachet Les Pucelles Premier Cru, but in the end chose the Chevalier-Montrachet for its sheer mouthful-for-mouthful joy, even if to be honest it is incredibly difficult to find. A true special-occasion bottle, this is a stunning wine fermented in oak casks and given the barest hint of filtration at bottling. Fleshy, grilled peaches dominate on the attack, but the stone fruits give way quickly to gunsmoke and pulsating touches of flint, as the wine retracts in on itself as it moves through the palate. It's going to run and run, that much is clear, and is possibly the clearest indication of the idea of terroir and the concept of 'minerality' in any wine I can imagine tasting. It's a testament to the brilliance of Anne-Claude.

'I had their Chevalier-Montrachet 2013 with a crab, truffle and porcini flan at Bouley and nearly wept!'
Jeff Harding, Sommelier, The Waverly Inn, New York

André Ostertag
Domaine Ostertag

87 RUE FINKWILLER, 67680 EPFIG, ALSACE, FRANCE
WWW.DOMAINE-OSTERTAG.FR

Many of us born in the 70s can relate to André Ostertag when he shares his teenage memories of marching against nuclear power. In André's case it was Fessenheim, the oldest plant in France that was located in his native Alsace.

'I had an early political awakening, and it took me quickly to a greater environmental awareness that became essential to my winemaking. I also have a strong memory from around the same time of visiting biodynamic gardens in Geisenheim and noting the force and vitality of the plants. I bought Steiner's book but didn't understand a word of it, so put it to one side, and only came back to it very slowly. I'm not political anymore, but the need to act on my beliefs has stayed with me.'

You're always going to get an honest answer from André. It's right there in his wines, which just burst with his pathological determination to avoid any bullshit. It might have been a progressive journey that took him back to biodynamics, but it was via one of France's most revered winemakers in the form of Dominique Lafon, who he became close friends with.

'Seeing the effect on his vines in Burgundy brought me back to the necessity to try it out for myself.'

In 1997 he returned to Alsace, began experimenting and within a year had converted the entire estate.

'The tough thing is putting to one side our need for rational explanations for everything. Steiner's writings can seem magical at times, and it can be frustrating. But once you have made the leap, everything changes. The powerful thing about biodynamics for me is its artistic side. There are pillars to follow but no recipes, so it's up to you to adapt it to your own piece of land with its own characteristics. What you will need to do is not the same as someone else. It's hard not to get a kick out of that.'
André Ostertag

DOMAINE OSTERTAG FRONHOLZ PINOT GRIS ALSACE 2014, FRANCE

DOMAINE OSTERTAG
VIN D'ALSACE
Appellation Alsace Contrôlée
2014
FRONHOLZ
Pinot Gris

Crayfish tails or a lovely fresh plate of ceviche is what comes to mind immediately here. These are rich but crystalline lemon and roasted-peach flavours that would be awesome with a plate of raw or super-fresh fish lightly doused with fragrant spices. Fairly subdued on the nose, this saves its firepower for a caressingly silky texture and a long finish. This could even work at the end of a meal if served with grilled peaches (I'm thinking on a barbecue) with some mint leaves. Brilliant quality, and an excellent way to see how Alsace can tease flavour, depth and interest out of the Pinot Gris grape.

DOMAINE MATASSA BLANC VIN DE PAYS DE CÔTES CATALANES 2014, FRANCE

WWW.MATASSAWINE.FR

Get excited about the potential of the Roussillon region of Southwest France with this blend of Grenache Gris and Macabeo, most of which are 120-year-old bush vines. South African winemaker Tom Lubbe moved to Calce in 2001 and has made increasingly striking wines ever since. This one is rich and round yet slaps you around the face with acidity, a turbo-jolt of freshness. It deserves to be carafed, because the flavours are changing constantly and continue to do so with aeration. It has the gulpability factor in spades, impressive at such a refreshingly low alcohol. Just add sushi.

DOMAINE TRAPET BEBLENHEIM RIESLING ALSACE 2013, FRANCE * *

WWW.DOMAINE-TRAPET.FR/EN

Trapet is one of the most famous names in Burgundy, and I can not recommend highly enough that you try their Gevrey-Chambertin wines (Clos Prieur 2012 is a particular favourite). This Riquewihr property was owned by Andrée Trapet's grandfather, and since 2002 she has run it with her husband Jean-Louis. They are committed to biodynamics on both estates. Her grandfather planted the vines here after the frost of 1956, and these were the 1.5 ha at the origin of the estate. Clay-dominant soils translates into rich, dense flavours, and with around 10g of residual sugar the acidity is so beautifully crisp it seems bone dry. Candied lemon shot through with quince, slate, gun-smoke. Will stand up to rich flavours, so do your worst. Salty pancetta could be perfect. Demeter-certified.

JOSMEYER BRAND GRAND CRU PINOT GRIS 2014, FRANCE

WWW.JOSMEYER.COM

Two sisters Céline and Isabelle run this Alsace property, which dates back to 1854, and make a range of amazing wines. It is their Pinot Gris from Brand Grand Cru that highlights the potential of this often-under-estimated grape. A blast of apricot fruit gets things rolling from the off, followed up by a rich and mouthfilling mid-palate with a seam of white flower freshness that keeps things on the up and up. This is a great wine to match with a fleshy fish like salmon or with spicy crab cakes.

LA BASTIDE BLANCHE BANDOL CUVÉE ESTAGNOL BANDOL 2015, FRANCE

WWW.BASTIDE-BLANCHE.FR

Rich yellow in colour, giving a hint of that wonderful Provence sunshine. Great structure and depth of fruit, relatively low in acidity and yet still loaded with fresh white flower aromatics that pick things up on the finish, this is a blend of Clairette, Ugni Blanc, Bourboulenc and Sauvigon Blanc. These age brilliantly, and their structure make them great lunch choices with a Niçoise or grilled chicken salad.

MICHEL AUTRAN CIEL ROUGE VOUVRAY 2014, FRANCE *

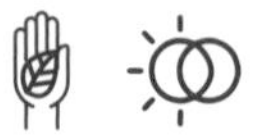

169 RUE DU 11 NOVEMBRE, 37210 NOIZAY

A former doctor who changed career to wine, Michel Autran is making Chenin Blanc wines that are tense, nervy, direct and super-charged. His Ciel Rouge bottling comes from flint and clay soils and is an amazing match to rock oysters or any shellfish. Just give it time in the glass, because those tight apple-and-pear flavours need some time to loosen up to show their glory and deepen into quince, beeswax and lemon cream.

OLIVIER HORIOT EN VALINGRAIN ROSÉ DE RICEYS 2012, FRANCE

WWW.HORIOT.FR

Olivier Horiot's property is located in the southernmost tip of the Champagne region, in the Côtes des Bars. He makes some excellent Coteaux Champenois but the real fascination here are the traditional still wines of the area known as Rosé de Riceys. For this, Olivier uses two separate plots of Pinot Noir, vinifying them separtately, starting with a small amount of foot treading before adding the rest of the bunches (whole). It is then aged for at least a year in oak barrels and bottled unfined and unfiltered with minimal sulphur. So – not your average rosé; certainly it is deeper in colour and flavour than most rosés. There is an austerity and savoury edge to the fruit, more redcurrant and grilled herbs than soft summer fruits. Don't overpower it – keep things simple with a beef carpaccio served with plenty of olive oil and freshly ground pepper.

WEINGUT DR. WEHRHEIM MANDELBERG PINOT BLANC PFALZ 2015, GERMANY

WWW.WEINGUT-WEHRHEIM.DE

This Pfalz estate has three generations of the same family still active in the business – grandfather Dr Heinz Wehrheim, father Karl-Heinz and son Franz are all making their contribution alongside various other family members. Their stunning Pinot Blanc is grown on limestone soils in the Mandelberg vineyard and has been named best white Pinot in Germany more times than I can count (alright, five). Mirabelle and pear, gorgeous freshness with some smoky edges. With extended lees contact after fermentation it makes a beautiful pairing either with a robust white fish or with cheese after a meal.

MEINKLANG GRAUPERT PINOT GRIS 2015, AUSTRIA ✱

WWW.MEINKLANG.AT

This stood out in a tasting of over 100 biodynamic, organic and natural wines, easily one of the most enjoyable there, and it has delivered again on several tastings back home, particularly paired with meaty fish like grilled sole or tuna steak. Apricot, nectarine, peaches. Basically, you name the stone fruit, it's in here. So juicy, with a hazelnut, smoky edging. Graupert means wild in the local dialect of Burgenland, which here refers to their vineyards, where the vines have been completely unpruned for over 15 years.

WEINGUT SEPP UND MARIA MUSTER GRÄFIN 2013, AUSTRIA

WWW.WEINGUTMUSTER.COM

An estate that dates back to at least 1727, today run by Sepp and Maria Muster and their three children. The extended (usually two weeks) maceration of the 100% Sauvignon Blanc grape on skins gives this wine its warm orange colour, and its rich, round taste. Expect a slightly nutty, oxidized edge to rich citrus fruits. Good match for spiced grilled vegetables.

GEYERHOF GRÜNER VELTLINER GUTSRESERVE 2008, AUSTRIA *

WWW.GEYERHOF.AT

Josef and Maria Maier make this rich and complex Gruner Veltliner, aged on the lees for seven years before release. A touch of residual sugar at 5.5g/l but it is so perfectly balanced by the acidity that the overall impression is of power and grace, never a hint of sweetness. A wine that stays with you – first through its persistency on tasting, and then that you just keep thinking about over the following few days. Ensure you pick a moment to enjoy this slowly over a good meal. White asparagus in a hollandaise sauce would be a good choice – certainly a popular one in Austria.

WEINGUT WERLITSCH EX VERO III 2012, AUSTRIA

WWW.WERLITSCH.COM

A dark, rich sunset yellow colour, this is utterly compelling from the first drop, with mandarin and slightly bitter blood-orange flavours set against white flowers, giving way to richer lemons. Ex Vero in Latin means 'from the true and genuine, totally real', and certainly this wine feels like it was made with care and heart by winemaker Ewald Tscheppe. These grapes come from the poorest soils of the property, meaning low yields given further concentration by sunny south-facing slopes. A great food wine, will easily hold up against a smoked fish, or spiced sweet-potato curry. Demeter-certified.

ANGIOLINO MAULE LA BIANCARA PICO VENETO IGT 2015, ITALY

WWW.ANGIOLINOMAULE.COM

A lovely singing minerality shines through this wine. Made with 100% Garganega, a local variety that Angiolino Maule strongly believes best expresses his terroir and philosophy. It is a fantastic match to a light saffron risotto. I just love the back story with these guys – Angiolino and his wife Rosamaria made pizza for years before buying a plot of land in the hills of Sorio di Gambellara where they planted vines to go alongside the olive and fruit trees (check out their homemade jams if you can). Today they work alongside their eldest sons Francesco and Allessandro.

COLOMBAIA VINO BIANCO TOSCANA IGT 2015, ITALY

WWW.COLOMBAIA.IT

In a tiny winery in the hills of northern Tuscany, winemaker Helena Variara makes skin-contact white with Trebbiano and Malvesia grapes. It has a lovely burnt orange colour and is almost like a fruity beer on the nose, with that touch of bitterness on the finish so typical of orange wines, which makes them so food friendly. There is a lovely unforced feeling to the mandarin notes, a beautiful sense of freshness and simplicity. Think spicy food, from pad Thai noodles to a green curry.

LA STOPPA AGENO BIANCO EMILIA IGT 2011, ITALY ✸

WWW.LASTOPPA.IT

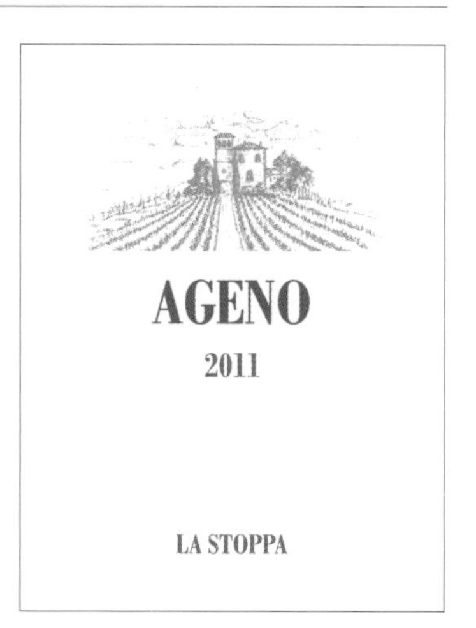

This is full-on natural wine – often pretty funky flavours are going on, and it's hard to miss the highly oxidative notes with honey, dried figs, orange rind, wax. And yet… it is just great, a total crowd pleaser that somehow beats all the odds of my uptight wine background to totally deliver on flavour and authenticity. And not just this one bottle – Elean Pantaleoni and the team at La Stoppa manage to deliver over and over again. The colour here comes from 30 days on the skins, from 60% Malvasia di Candia Aromatica, 40% Trebbiano and Ortrugo grapes, aged half in oak and half in stainless steel. Picking an equally unusual food pairing – this might even stand up to artichokes.

Above: The natural vineyards of La Stoppa in Italy.
Overleaf: Meinklang winery in Austria also rears cattle.

Saša Radikon
Radikon Oslavia

LOCALITÀ TRE BUCHI, N. 4, 34071 GORIZIA, ITALY
WWW.RADIKON.IT/EN

The name Radikon stands loud and proud over orange wines, the godfather of them all, or in Italy at least. Since Stanko Radikon reintroduced this traditional winemaking style in 1995, they have slowly but steadily built a cult following to rival the most iconic estates in the world.

'Skin-contact winemaking was traditional to the Friuli region,' says Saša Radikon, who today has taken over from his father Stanko. 'It was all lost in the 70s and the 80s because the fashion turned to technically made wines, with oenologists and new equipment. Everything had to be filtered and clean, and the old style of wine was rejected. They lost the tradition until my father and Josko Gravner reconnected to it.'

'I studied oenology with all of its emphasis on modern winemaking techniques, and my father used to say, "you are studying what not to do!" 'But it was always clear to me the style of wines that I wanted to make. I grew up here watching him in the winery, and when we started working together we discussed these new methods, of course, but we always liked the same style. It occurred to me recently that when he first launched the wines with just one-week maceration back in 1995, I was too young to experience them. Over time he moved to more extreme orange wines with two to four months of skin-contact, and I wanted to experience for myself what it was like in the beginning, in the style my grandfather made.'

'I have now introduced a range of lighter, more fruit-driven wines. It means that after ten years full-time at the winery, I know there is a piece of me also in these wines.'
Saša Radikon

RADIKON OSLAVIA, RIBOLLA GIALLA VENEZIA GIULIA IGP 2003, ITALY

It's questionable whether orange wine would really have found its international fame without Stanko Radikon, a wine legend from the Friuli region of Italy who died in 2016 but whose reputation lives on. The Ribolla Gialla grape gets two to four months skin-contact, giving it a rich amber colour, and the wine is kept back for at least three years before heading out to market. The main impression is of complexity – you'll find bitter mandarin oranges set again white truffles, caramelized oak, mint and rich lemon. Saša Radikon recommends Japanese home-style foods like blackened cod or Teriyaki salmon or Teppanyaki beef.

TENUTA TERRE DELLE TERRE DOC ETNA ROSATO 2014, ITALY

WWW.TENUTATERRENERE.COM

Rich salmon in colour, abundant in grilled sage, heather and thyme on the nose. Dry in style, the red fruits are tucked in rather than overly generous. Made from a grape indigenous to Sicily called Nerello Mascalese, there is a beautiful brightness to this rosato, and it's brimming with personality. The estate gets 80% of its electricity needs from solar panels, and (almost all of) the rest from burning the cuttings from 1600 olive and other trees on the property. When it comes to food pairing, they recommend prosciutto, figs and cantaloupe. Who am I to disagree.

PHEASANT'S TEARS CHINURI AMBER WINE 2015, GEORGIA

WWW.PHEASANTSTEARS.COM

Grown in organic ('since forever, but not certified' is how winemaker and co-founder John Wurdeman puts it) vineyards at 550m (1805ft) above sea level in the Alazani Valley. The resulting grapes are left on their skins for three weeks to get the beautiful amber colour, then fermented and stored in the traditional Georgian clay pots known as *qvevri* for nine months. Some of the Pheasant's Tears *qvevri* date back to the 19th century, so you really are drinking history. Adding to the sense of mystique is that Chinuri is an ancient grape indigenous to the region. It is stunningly fresh and light, once you have given yourself a few moments to adjust to the honeysuckle and slight oxidized notes on the attack. A beautiful wine, the essence of the idea of handcrafted.

'This wine needs food and really cannot be understood without it. I would recommend a roast quail with verjus, and pomegranate.'
Franck Moreau MS, Group Sommelier, Merivale, Sydney

Above: Radikon Oslavia sits at the Slovenian border in northern Italy.

BINDI WINES KOSTAS RIND CHARDONNAY MACEDON RANGES 2015, AUSTRALIA

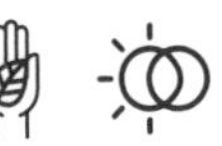

WWW.BINDIWINES.COM.AU

This manages to balance the swagger of an oaked Australian Chardonnay with a tempering sense of gentle uplift through the body. Grown on quartz-heavy volcanic soils, these low-yielding vines see natural fermentation and eight months in barrel on the lees. A mouthfiller, complex with slightly candied lemon playing off a tight seam of lime. Would be great with turbot or butternut squash curry.

CULLEN AMBER WINE MARGARET RIVER 2014, AUSTRALIA

WWW.CULLENWINES.COM.AU

You can take a biodynamic garden tour at this winery and go for full immersion of the benefits at their on-site restaurant. This wine is a blend of Sauvignon Blanc and Semillon, vinified in a mix of tanks and amphora. Beautifully nuanced, tangerine and candied ginger abound. Owner Vanya Cullen also offsets all carbon emissions through programmes such as carbon sinks and planting of native trees and shrubs and through a large reforestation project in the Yarra.

'Not completely a classic orange wine but nevertheless with some skin-contact and as a result an attractive copper colour with great freshness on the palate. Sea bass cooked in a sea-salt crust would make for a formidable combination.'
Gerard Basset MW, Best Sommelier of the World 2010

MARCHAND&BURCH
FINE WINES OF WESTERN AUSTRALIA AND BURGUNDY

MARCHAND AND BURCH PORONGURUP CHARDONNAY MARGARET RIVER 2015, AUSTRALIA ✱ ✱

WWW.BURCHFAMILYWINES.COM.AU

Burgundian winemaker (and as they put it 'biodynamic ambassador') Pascal Marchand has in his time managed Domaines Comte Armand and de la Vougeraie. Here he has teamed up with Jeff Burch of Howard Park and MadFish Wines. This is serious grown-up Chardonnay, from some of the oldest vines in the Porongurup sub-region of the Great Southern region – itself a volcanic outcrop whose soils help to keep yields naturally low and flavours intense. It will age beautifully, but already is singing with its emphasis on white pear, jasmine and steely citrus. It emits an energy that is unmistakable from the first sip. Delicious with a fresh mango and avocado salad.

Jaysen Collins
Massena

CNR LIGHT PASS AND MAGNOLIA ROADS,
VINE VALE, SA 5352, AUSTRALIA
WWW.MASSENA.COM.AU

I can't tell you how many people who helped me research this book recommended Jaysen Collins. His wines are burning up Australia, as, not too slowly, he creates a reputation as a passionate, committed winemaker. Not bad going for someone who was born in Barossa but studied for a business degree at the University of Adelaide before becoming a qualified accountant. This took him to management roles in various wineries, before he got the itch for making his own wines – at first by maxing out his credit cards to buy grapes with his friend Dan Standish.

'The defining moment for me to change was when I was working at St Hallett winery in the accounting office and I attended a breakfast as a guest of our bank. The speaker was author Bryce Courtney talking about having the courage to do things you've always wanted to do. I walked directly out of the breakfast and booked a ticket to Europe, travelled through the majority of the wine regions and returned to the Barossa and started Massena.'
Jaysen Collins

Slowly but surely over recent years, he's moved towards organics for the fruit that he grows and buys (from around eight vineyards across Barossa).

'My own vineyard didn't receive any sprays last growing season and the ground is not worked, I encourage the native rye grass to grow under vine, as it dies off in the growing season and helps retain any moisture in the soil. In the vineyards I don't own, I work with the grower in the sections I get to ensure they are best for winemaking. I use harvest dates to help reduce the need for acid additions, and plant varieties such as Saperavi, as it has natural acidity and tannins, so in a warm climate it takes away the need for acid additions.'

MASSENA THE SURLY MUSE
BAROSSA VALLEY 2016, AUSTRALIA

Old school friends turned winemakers, Jaysen Collins and Dan Standish source grapes from various growers across the Barossa for this blend of 75% Viognier, 25% Marsanne. The Viognier is sourced from a single vineyard in Koonunga Hill, one of the coolest regions of Barossa, that has sulphur and copper sprays and tilling to manage the weeds. In the cellar, Viognier is fermented in tank and Marsanne in barrel, both with native yeasts, and then aged in French oak. Touches of caramelized apple, ginger and sumac spice complement the fresh acidity that runs through this wine and balance the weight of ripe lemon and apricot. Perfect with a warm potato salad with olive and anchovy dressing.

NGERINGA CHARDONNAY ADELAIDE HILLS 2014, AUSTRALIA

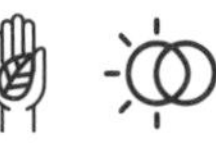

WWW.NGERINGA.COM

This has a totally seductive combination of ripe fleshy fruit with a sense of vigour and energy that drags you forward through the palate. Stone fruits of nectarine and apricot against tart lemon, the whole thing deepens and gathers interest in the glass like all great Chardonnay. Erinn and Janet Klein created Ngeringa wines in soils that were once part of the Jurlique herb farm and were among the very early adopters of biodynamics in Australia. Alongside the vines they have cattle, sheep, orchards and an olive grove.

'Medium-bodied wine with very good acidity on the finish, makes me want to eat crayfish with kombu butter or simply a lobster roll.'
Franck Moreau MS, Group Sommelier, Merivale, Sydney

SORRENBERG CHARDONNAY BEECHWORTH 2015, AUSTRALIA * * *

WWW.SORRENBERG.COM

This has a waxy edge to it, more Puligny-Montrachet than Chablis, with lashings of pretty lemon curd laid over a steely core and pin pricks of reduction that unfurl in the glass, revealing apricot and nectarine richness without ever straying into the clichéd buttery Chardonnay. Mouthwatering and full-bodied, go for chicken livers or wild mushroom risotto with this. The estate is also well known for its excellent Gamays. Demeter-certified.

MILLTON VINEYARDS TE ARAI VINEYARD CHENIN BLANC GISBOURNE 2015, NEW ZEALAND

WWW.MILLTON.CO.NZ

A wine that will make you fall in love with Chenin Blanc, underlining its ability to be both steely and strict while still having rich seams of citrus, pear and honeysuckle that seem to envelop your mouth. There is a touch of residual sugar (around 6g/l) that translates into a richness in the mid-palate but this tightens its grip beautifully towards the end. Chinese pork with soy, sesame, garlic, ginger and honey will work great.

Jacques Lurton
The Islander

THE ISLANDER, ESTATE VINEYARDS, 639 BARK HUT RD,
CASSINI, KANGAROO ISLAND, SA 5223, AUSTRALIA
WWW.IEV.COM.AU/IEV

Sometimes I like to imagine the sprawling Lurton clan sat around the supper table arguing about winemaking. There are three generations active in various parts of the business, 15 family members and counting, with over 30 vineyards worldwide and 1300 ha of vines between them. And a good proportion of that is being farmed biodynamically. Bordeaux is their home base, where you find Bérénice at already-biodynamic Château Climens in Sauternes (see p.231) alongside her brother Gonzague and his wife Claire. Another brother, Henri, is edging that way.

Jacques represents another branch of the family – cousin to Bérénice and Gonzague – and he has taken the idea of organics and biodynamics over to Australia; while his brother François does the same in the Languedoc, Spain, Portugal, Chile and Argentina. Which brings me back to the dinner table. In the early years, Jacques' wines were often big and broad, but today they are sculpted, crafted and increasingly made with either low or no sulphur.

'I walked away from blockbusters. Today my palate takes me more naturally to Burgundy or northern Italy – places with natural acidity. And although we have reached a point where oenological products are extremely efficient, and we can make wines smoothly and efficiently by using them, it increasingly feels essential to make as little intervention as possible. When I was younger the winemaking took over. But now it's the vineyards that count, and being as honest to their real nature as possible. It took me being my own boss, and being small enough in terms of production levels, to have the freedom to take the risk.'
Jacques Lurton

THE ISLANDER
estate vineyards

THE ISLANDER WALLY WHITE KANGAROO ISLAND 2013, AUSTRALIA

The vines for this wine come from Kangaroo Island in South Australia, off the coast from Adelaide. The flinty character hits you straight away, startlingly so considering its blend includes old-vine Semillon, then on the mid-palate it softens to reveal creamy, apricot-edged flesh. It is a wine that will age beautifully, with lovely natural acidity and great balance. I just love it – and even more so with a plate of Serrano ham.

03

Light & Sculpted Reds

Pascaline Lepeltier MS

Master Sommelier at Rouge Tomate, Chelsea, New York

Right from my first job as a sommelier in the Loire Valley, France, back in 2005, I have been drawn to authentic, natural wines. I love their energy, their honest commitment to integrity and to farming in a way that is fair and sustainable. These are the wines that I drink myself, and are often fantastic for both complexity and value.

This method of artisan winemaking, with its focus on vibrant flavours, works perfectly with fresh, sculpted reds – a broad category that is sometimes considered to be 'light' – as meaning little or simple and easy wines. This is too bad as this category offers some of the most complex wines in the world, as well as fantastic bottles to pair with a very large array of dishes and ingredients. It can instead be useful to think of them as the red counterpart to the full-bodied, skin-contact whites.

The lifted, lighter aspect of the wines can be understood in three ways. The wine is either:

light in alcohol – when the wine is produced in a cooler climate areas like northern France (Loire Valley, Coteaux Champenois, Alsace, Burgundy, Beaujolais), Switzerland, northern Italy, Spain's Galicia, Germany, Styria, Ontario, New York State, Oregon, coastal California, Adelaide Hills, Yarra Valley, some higher elevations vineyard in Chile and so on;

light in tannins – here either the grape is naturally light in tannin or the winemaking is conducted to extract very little (such as carbonic maceration in the Beaujolais, or gentle infusion of the skins during maceration without excess movement, eg. punching down or pumping over, that increases the extraction of the tannin in the skin). The most notable grapes in this style are Pinot Noir, Gamay, Grenache, Pineau d'Aunis, Grolleau, Pelaverga, Blaufränkisch, Mencía, Blauer Wildbacher, Frappato, among others;

light in both – when grapes are grown in a cooler environment and treated lightly in the cellar.

In all cases, the acidity is quite crucial, as it will sculpt the body of the wine. These wines can be great with meat, both red and white flesh, including sweetbreads. But when you think about the freshness of the wine's structure, the amount of possible pairings opens up, as you can serve these wines at different temperatures, including slightly chilled, to change their impact. They then become your best weapon when you have to handle tricky food matches.

This is the category of wines that can go with fish, of course. Grilled Mediterranean fish works wonderfully with the Frappato grape, planted primarily in Sicily. Or try river fish with a butter sauce paired with Loire Cabernet Franc made with carbonic maceration. Or Hawaiian walu tuna with a lightly chilled Grenache. These wines also handle vegetables extremely well, including the difficult category of bitter greens. Mushroom tartare with Worcestershire sauce can be perfect with a Pinot Noir from Mâcon. Another great pairing is a kale Caesar salad with a Piedmont Pelaverga or Freisa. Or a sautéed dish of spring vegetables – fresh peas, baby carrot, green asparagus, garlic and onions – with Loire Grolleau.

If you want to drink a red with spicy food, these fresher styles will also be your best option, you need to pick a wine with low tannin and alcohol as both of these elements tend to enhance the burn of the spice. The Styrian Blauer Wildbacher from Austria is often a great option.

Sculpted, fresh reds are also a better option with cheese, especially the lighter tannins, as they won't bind the proteins and won't leave too much of a metallic taste. For wash-rind cheese such as taleggio from Italy or Gruyère from France, Pineau d'Aunis can be a great option with its forest floor aromatics.

Pascaline's Wine Picks

BENOIT COURAULT, LA COULÉE, ANJOU, LOIRE, FRANCE

DOMAINE DE BELLIVIÈRE, ROUGE-GORGE, COTEAUX DU LOIR, FRANCE

FRANZ STROHMEIER FRIZZANTE (BLAUER WILDBACHER), STEIERMARK, AUSTRIA

GUISEPPE RINALDI, BAROLO, PIEDMONT, ITALY

PRINCIPIANO FERDINANDO CHILA FREISA LANGHE, PIEDMONT, ITALY

LIGHTFOOT & WOLFVILLE VINEYARDS ANCIENNE PINOT NOIR NOVA SCOTIA 2014, CANADA

WWW.LIGHTFOOTANDWOLFVILLEWINES.COM

Winemaker Josh Horton is absolutely pushing the boundaries of what has been achieved in Nova Scotia to date. Located in the Annapolis Valley, this estate was founded in 2009 (although the property had been in Mike Lightfoot's family for generations and is a working farm). It focuses on Chardonnay, Pinot Noir and Riesling – all grapes that respond well to a cooler climate – and has already built up a good reputation even though it only began releasing wines in 2015. Josh uses wild yeast fermentation and French oak barrels for ageing. Pair this with a wild game pie or grilled porcini mushrooms on toast, and enjoy the earthy softness of the wild cherry fruits.

AMBYTH ESTATE MAIESTAS PASO ROBLES 2012, USA ✸ ✸

WWW.AMBYTHESTATE.COM

A dry-farmed estate in California's Paso Robles region, with vines, olive and fruit trees, bees, chickens and dairy cows. Grapes are crushed by foot, fermentation is with natural yeast, no filtering or fining and zero additives. A little touch of vanilla oak gives sweetness, the wines are packed with ripe red fruits and personality: perfect for a summery lunch of grilled chicken salad with quinoa, fresh mint, parsley and coriander. An energy-filled blend of 60% Syrah, 22% Grenache, 14% Mourvèdre, 4% Counoise, these Rhône varieties are given full justice. Demeter-certified.

Above / Overleaf: A new generation of biodynamic wines from Lightfoot & Wolfville in Nova Scotia, with a focus on aromatic varieties that suit the climate.

Cathy Corison
Corison

PO BOX 427, ST HELENA, CA 94574, USA
WWW.CORISON.COM

Cathy Corison is one of the key winemakers who inspired me to write this book. Her wines are the absolute encapsulation for me of sculptured, supple, elegant Cabernets, and the fact she never waivered from this style while so many around her in Napa were pushing new oak and alcohol levels ever further and higher makes her something of a personal hero.

Her wines always have the power and density that you expect from Napa, but they combine a delicacy and a sense of place that make you smile. Cathy has a UC Davis Masters in Oenology and spent many years working for big names like Staglin and Chappellet. Today she makes her own wines, some from sourced grapes but mostly from her own. Her Cabernets are some of the oldest in Napa and only replaced when truly necessary – something that she suggests is crucial for a wine's complexity. 'Cabernet is always powerful, but the interest for me is the intersection of power and elegance. I believe that Napa is perfect for this style, as we have the fog and the cool nights.'

'It seems almost a moral obligation in this part of the world to make wines with a life, because you can't do that everywhere, it's a gift.'
Cathy Corison

In addition to the 100% organic wine selected here, Corison also makes a Napa Valley Cabernet from sourced grapes, 'that inches closer to organic every year, but it's not there yet,' she says, direct as ever. 'In fact, a bit over a year ago we had the great good fortune to purchase one of the three vineyards I've sourced for the Napa Cabernet for nearly thirty years. We immediately began farming it organically. Our Kronos vineyard, on the other hand, has been farmed organically for twenty years.'

CORISON KRONOS NAPA VALLEY 2013, USA

Cathy Corison makes her wines to showcase balance and vibrancy and Kronos is the place where she first worked on moderating ripeness in the fruit back in the 80s. In winter, it is a sea of yellow, with wild mustard between the old-vine Cabernet Sauvignon fruit. Naturally low yielding and concentrated, it opens up beautifully and the fragrance entices you towards it. This wine majors on blueberry and black cherry fruits, with an electric core running through it.

'I would happily serve this with a variety of hearty meat-based dishes, especially duck (ideally in a green peppercorn sauce) or a perfectly dry-aged ribeye.'
Kelli White MS, Press Restaurant, Napa

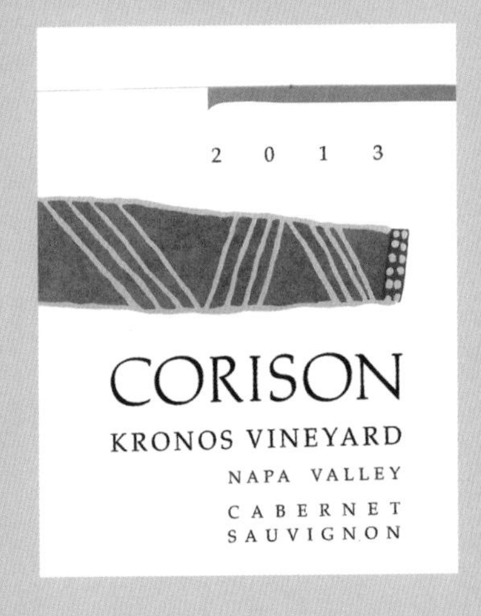

Veronique Drouhin

Domaine Drouhin

PO BOX 700, DUNDEE, OR 97115, USA
WWW.DOMAINEDROUHIN.COM/EN

Calm, capable and unflappable, Veronique Drouhin always looks perfectly elegant and in control to me – particularly impressive considering she is chief winemaker for her family's renowned Burgundy house and also for the wonderful Domaine Drouhin in the Dundee Hills in Oregon's Willamette Valley, working alongside her three brothers Philippe, Frederic and Laurent. They are the fourth generation of the family, with eight Drouhin children promising the fifth.

A visit to the Drouhin's 13th century cellars under the streets of Beaune is one of the best visits you can do in Burgundy, and the family's wines on both sides of the Atlantic are truly exceptional. They work entirely bioynamically in Burgundy, and are in conversion in Oregon, with no chemical pesticides or insecticides and with a significant part of the vines already in biodynamics. Her brother, and vineyard manager, Philippe says, 'we became conscious that there were no answers to a very simple question – what becomes of the chemicals that are applied to the vines? I didn't have good answers, so I decided to return to the way of farming that had been done for centuries. It was not called organic but it was organic. At first it was not a good marketing argument; that has only come recently. My reason is technical, and to better express the terroir.'

Veronique has worked at the family firm since 1986, and has been a key part of the change to organics since 1990, which was instigated by Philippe. She says the pleasure of working between the two countries is the challenge and the exchange between two cultures, plus the expression of their Pinot Noir.

'The soil in Burgundy is limestone with high mineral content, which gives a natural elegance to the wine. In Oregon most soil is volcanic and it's very different to work with. You have to be careful not to overdo things. I have learned that you can go too far with extraction, or with new oak, and it can mask the character of the Pinot. We celebrate 30 years in Oregon in 2017, so we have been through many different styles of vintage, but it is fascinating how we are always learning something new.'

Veronique Drouhin

DOMAINE DROUHIN LAURÈNE PINOT NOIR DUNDEE HILLS OREGON 2011, USA * * *

Drouhin is a name that goes a long way in Burgundy – in fact all the way back to 1880. Robert Drouhin bought in Oregon in the 80s. For my money this Laurène bottling is among the best examples of Oregon Pinot that you can hope to find. Liquorice and blueberry abounds, you can just close your eyes and imagine your lips getting stained with the best produce of summer. There is true integrity to the fruit, fresh but layered and with just the right amount of yielding to pressure when needed. Soft and long tannins, it gives an instant hit of pleasure and yet spins itself out perfectly, proving an ideal match for seared scallops with truffle oil.

EVENING LAND SEVEN SPRINGS VINEYARD PINOT NOIR 2014, USA

WWW.EVENINGLANDVINEYARDS.COM

This supremely elegant Pinot Noir from the Eola-Amity Hills area of Oregon is vinified with 25% whole-cluster fruit in open-top concrete and oak vats. The vines in the oldest blocks in the vineyard are 30 years old, dry-farmed and planted by Al MacDonald in 1984. The resulting fruit is pretty much the ideal expression of a warmer, fruit-driven Pinot that combines plush blueberry fruit with an austere touch of darker spicy notes and firm tannins. There is a feeling of excitement around this property, founded by film producer Mark Tarlov but now owned by award-winning author and sommelier Raj Parr, together with winemaker Sashi Moorman – with consultancy from Dominique Lafon. Grilled pork chops and red cabbage would work well here.

FORLORN HOPE TROUSSEAU NOIR 2015, USA

WWW.FORLORNHOPEWINES.COM

Trousseau is an unusual grape that not many people are working with. This wine gives a sweet hit at first, and is soft and pale in colour. So there you are feeling all relaxed and easy, then it kicks in with a grip of tight tannins. After a few minutes in the glass, it has evolved to an array of grilled herbs and savoury fruits, and would be a flexible food wine – it could stand up to a game pie, but also the cheese board. Winemaker Matthew Rorick grows the Trousseau on schist and limestone soils at Rorick vineyard in Calaveras County. It's a revival, as he puts it, of the pre-Prohibition traditions: hand-picked, whole-cluster ferment in open-top vats and spontaneous fermentation. Matthew is worth following for his collaborations with winemakers around the world, from South Africa to New Zealand.

KELLEY FOX MOMTAZI VINEYARD PINOT NOIR OREGON 2015, USA

WWW.KELLEYFOXWINES.COM

The definition of an artisan winemaker, Kelley Fox makes fresh, vibrant wines that are impossible not to fall in love with. There is a masculine, stony edge to the 100% Pinot that teases you a little, as it forces you to takes things slowly. Give it time to open before the soft redcurrant and rose-petal aromatics become clear. Kelley is attentive to her vines throughout the year and yields are low. Grown on volcanic soils, this has a vibrancy, an effortless sense of extracting the heart of the fruit. Try with baked crab (no heavy sauce). Demeter-certified.

Randall Grahm
Bonny Doon Vineyard

328 INGALLS STREET, SANTA CRUZ, CA 95060, USA
WWW.BONNYDOONVINEYARD.COM

Randall Grahm is a man not easily contained in a short profile. California's resident maverick winemaker for over 30 years, it was surely always only a matter of time until the man behind Bonny Doon found his way to biodynamic winemaking. Certainly he's never been afraid of doing things because he believes in them, and particularly when others are telling him that it will never work.

> 'If my philosophy could be summed up it would be that wines of place are the only ones that truly matter. And after many years of talking about this, Popelouchum has given me the opportunity to reflect that sensibility and those values.'

But Randall is a pragmatist as well as a dreamer.

> 'I think biodynamic farming is incredibly useful in bringing vitality to vineyard soils, but other criteria must also be met to produce truly distinctive wines of place. My allegiance is to the site and its optimal expression rather than to a particular ideology or practice.'
> *Randall Grahm*

RANDALL GRAHM POPELOUCHUM GRENACHE 2015, USA

Okay, so this particular wine doesn't have a track record, as 2015 is the first vintage, but the winemaker certainly does. Popelouchum is Randall Grahm's new vineyard project in San Juan Bautista, California. He is best known for his Bonny Doon wines and bought this property with the express aim of making biodynamic, terroir-driven wines that just might also solve some of the problems faced by winemakers experiencing climate change. There are numerous experiments going on, from the use of a charcoal substance called biochar to increase water retention in soils, to extensive growing and crossing of hybrid grapes to develop drought-resistant varieties. This Grenache comes from vine cuttings from Rhône icon Château Rayas, and has an exceptionally beautiful varietal character. Its flavours are pure and elegant, with gentle touches of liquorice root, crushed puréed raspberries, white pepper and extremely gentle oak. Try a caramelized onion tart with this.

DE MARTINO CINSAULT VIEJAS TINAJAS ITATA 2014, CHILE

WWW.DEMARTINO.CL

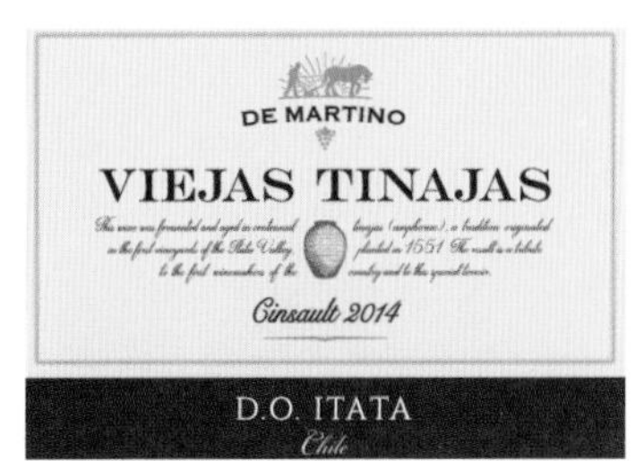

The De Martino family, today run by Pietro, Marco and Remo with their sons Antonio and Sebastian, own their own vines and source from dozens of growers around Chile. Since 2011 there have been big changes, including not using new-oak barrels for ageing and ensuring minimal intervention at every stage. The Cinsault vineyards are centred around Itata (sold under the Secano Interior appellation). Dry-farmed, ungrafted bush vines vinified with natural yeasts from uncrushed berries using earthenware jars called *tinajas*: lovely herbal, floral, spicy style (a lack of fruit descriptors is deliberate, because they are far from the dominant flavours of this wine although there is plenty of sour cherry). Earthy and slightly musky, in a good way, the tannins are so soft that this would be a perfect summer red to accompany grilled fish on a barbecue, and could easily be served slightly cool.

BODEGA CHACRA TREINTA Y DOS PINOT NOIR PATAGONIA 2015, ARGENTINA * *

WWW.BODEGACHACRA.COM

BOTTLE Nº. ______ OF 4050
CHACRA
PATAGONIA ARGENTINA
- TREINTA Y DOS -
2015 PINOT NOIR
ELEVE ET PRODUIT PAR PIERO INCISA DELLA ROCCHETTA

Located more than 1000 km (621 miles) to the south of Buenos Aires in Patagonia's Río Negro, this once-abandoned property has been brought back to life by Piero Incisa Della Roccheta of the rather more famous Tenuta San Guido estate in Bolgheri. They have managed to craft wines that are a million miles away from Argentinian block-busters – for a start because they focus on Pinot Noir, which is hardly a well-known variety here. The cool climate of Río Negro helps, and its isolation means the vines are all ungrafted (reported to help purity of flavours). The vines themselves date right back to 1932 and produce tiny amounts of fruit. Fresh and graceful with a marked berry character and firm tannins that could do with a good few hours in a carafe before drinking. They also do an excellent no-added-sulphur Pinot Noir called Zin Azufre. Plenty to play with here in terms of food matching, but don't overpower these delicate flavours.

FAMILIA CECCHIN GRACIANA 2014, ARGENTINA

RUTA 60 S/N, 5517 RUSSEL, MAIPÚ, MENDOZA

Various generations of the Cecchin family have been living in Argentina since emigrating from Italy in the 19th century. Today they have 70 ha spread over three plots. This is whole-bunch vinification, completely unoaked and bottled with no added sulphur. A beautifully fragrant Graciana, light, graceful, vibrant, with delicate red fruits. It's a grape with naturally high acidity, but the heat of Argentina translates this into lyricism rather than anything biting. Pair with a simple lunch such as a lentil and halloumi salad.

Maria José López de Heredia

Bodega López de Heredia

AVDA. DE VIZCAYA, 3, 26200 HARO, LA RIOJA, SPAIN
WWW.LOPEZDEHEREDIA.COM

Want to learn about traditional, natural winemaking techniques in the gentlest and most fascinating way possible? Pull up a chair with Marie José Lopez de Heredia. Or even better visit her stunningly atmospheric property in Haro, capital of Rioja Alta.

Time seems to stand still here. The 5000-litre oak vats are the same ones Maria José used to play hide and seek around as a child. The flagship Gran Reserva wine is bottled and corked by hand, then sealed under wax. And when they wash the barrels after ageing the wine, they use water from their own fountain.

'I believe being organic is an obligation, not an extra,' she says. 'Any owner of a vineyard should learn to respect nature. It comes with the territory, and it's part of a culture of respect for our consumers. If some years you get less crop, that's the risk. But it's one you have to be willing to take. My father believed in absolutely no intervention in the cellar. When I studied winemaking we were routinely told to acidify, but today I know that using the traditional grapes of Gracian and Manuelo have the same effect – they freshen up a blend naturally. We use only natural yeasts, we use tiny quantities of sulphur even though we age our wines for six years or more, but we believe good grapes have a natural protection against oxidation.'

'My grandfather and father followed this path before me. To make wine like this needs commitment, even economic sacrifice at times, but I want people to enjoy our wine and for me that is only compatible with following the traditions that have always worked for us.'
Maria José López de Heredia

BODEGAS LÓPEZ DE HEREDIA VIÑA TONDONIA RESERVA RIOJA 2004, SPAIN ✱ ✱

Maria José Lopez de Heredia describes her winemaking as 'more natural than natural wines'. The reds are made with no temperature control, beside open windows for a bit of ventilation, then aged in underground cellars where 20 years worth of vintages are stacked five-barrels-high in cobwebbed tunnels because they like to age their wines well before release. This is the flagship, an excellent vintage, tasting beautiful right now. From a blend of Tempranillo (75%), Garnacho (15%), Graciano and Mazuelo (10%), you get a light, fresh texture that stretches out endlessly in front of you. Dried red berries, soft damsons and creamy figs mix with touches of vanilla root and woody, earthy notes. I'll just warn you that there can be bottle variation – although as with burgundy, this can be part of its charm. Their own suggested pairing is with Welsh lamb stuffed with wild garlic, rosemary and anchovies. Or...

'Lightly smoked saddle of roe deer with grilled celeriac, mushroom jus and black truffle.'
Fredrik Lindors, Sommelier, Grand Hotel, Stockholm

Above: López de Heredia has one of the most traditional and beautiful cellars in Rioja.

UVA DE VIDA LATITUD 40 JOVEN 2015, SPAIN

WWW.UVADEVIDA.COM

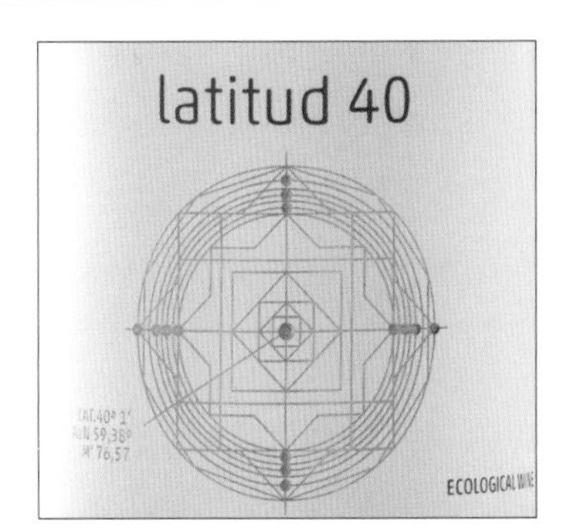

Carmen López and Luis Ruiz keep things simple at this estate, with 100% Graciano and only a few different bottlings. Graciano is a grape that has lost popularity in Toledo in recent years, just as it has in Rioja, so it is thrilling to see it being treated so well here, as it can be a totally lyrical experience when the winemaker gets it right. They bottle a Crianza version, which is slightly more structured, with a sprinkling of cocoa and soft-cured leather, but this Joven is almost impossibly fresh with strawberries and roses that highlight the floral, fresh side to Graciano. A perfect wine to accompany the simplest of grazing plates: hams, cheeses, olives, bread and olive oil.

BENOIT COURAULT LA COULÉE VIN DE FRANCE 2014, FRANCE

VALLET, 49380 FAYE D'ANJOU

Another young winemaker who is busy putting Loire on sommeliers' radars world over. Benoit is the son of a horse breeder who became passionate about wine (working at one point with natural wine legend Eric Pfifferling at Domaine l'Anglore in the Rhône) and found himself a couple of hectares of vines. They are located just outside the Anjou appellation (probably why he could afford them), but on excellent limestone soils. As a horse breeder's son, it's no surprise that he works part of his vines with horses, and all of them organically. His La Coulée is made with the Grolleau grape, 40-year-old vines, aged for just under ten months in barrel (no new oak). Totally succulent, pure blackberry and wild strawberry fruit, you can practically feel it bursting out of the glass. No added sulphur. And great value.

'Perfect paired with a rabbit and pistachio terrine with a salad of mache and halzenut.'
Pascaline Lepeltier MS, Rouge Tomate, New York

CHÂTEAU DES BACHELARDS COMTESSE DE VAZEILLES FLEURIE IGP COMTÉS RHODANIENS 2014, FRANCE

WWW.BACHELARDS.COM

Alexandra de Vazeilles (the countess) is part of a small group of dedicated producers focused on bringing out the best of the Beaujolais crus terroirs. Since buying the estate in 2007, Alexandra converted to organics and lavished attention on the vines. Beautiful, soft rose petals on the nose, then a punch of vibrant and just-sweet-enough red fruit, with the zing and energy of Gamay made in the right hands, and the grip from Fleurie's granite soils. Steak tartare would set this off nicely.

Above: Applying biodynamic preparations at Clos Puy Arnaud.

CHÂTEAU DU CÈDRE CAHORS CUVÉE EXTRA LIBRE 2015, FRANCE

WWW.CHATEAUDUCEDRE.COM

Pascale and Jean-Marc Verhaeghe produce some of the most nuanced and yet old school wines of Cahors. A blend of 97% Malbec and 3% Merlot, aged for 12 months in large oak casks and no added sulphur, it is a seriously moreish wine, intense and spicy with fresh dark fruits. It just gets better and better in the glass, I was totally gripped by the definition and vigour there is to the fruit here, it's luscious. Worth noting that the the rest of the wines made by du Cèdre are also organic, but are not 'natural' as they do use sulphur at bottling. Fresher than many Cahor wines, great with sautéed mushrooms or braised lamb.

CLOS DE TRIAS VENTOUX VIEILLES VIGNES 2015, FRANCE

WWW.CLOSDETRIAS.COM

A rich blend of 96% Grenache and 4% Syrah for this gorgeous Rhône wine. Run by Even Bakke, the vineyards are certified organic but not yet biodynamic ('though fully entrenched in Steiner's beautiful work' is how Even puts it). Even was born in Boulder, Colarado and raised in Norway; he founded Clos de Trias in 2007 after working for over a decade in California – building up plenty of influences along the way. Trias sits on the western foothills of Mount Ventoux, which takes it up to 500m (1640ft) in altitude at some points. The 'old vines' in this particular bottling are up to 80 years old, giving a richly powerful wine loaded with strikingly pure black fruit flavours. Match with pulses: black bean *tostadas*, with lashings of cumin and coriander.

Thierry Valette
Clos Puy Arnaud

7 PUY ARNAUD, 33350 BELVÈS-DE-CASTILLON, FRANCE
WWW.CLOSPUYARNAUD.COM

Castillon-Côtes de Bordeaux is tucked away in the far eastern part of the Bordeaux region as it heads towards the Dordogne and Périgord. It is hardly the best-known appellation in Bordeaux, and although it shares much of the same limestone soils as neighbouring St-Émilion, it seems to have missed out on its international renown. But, it does have one thing seriously going for it; 25% of Castillon wines are either farmed organically or biodynamically, giving it the green crown of Bordeaux, helped perhaps by the number of smallholders who live on their estates and so are there every day to focus on the labour-intensive work of organic farming.

Thierry Valette has worked in both areas. His family was long-time owner of the iconic Châeaux Pavie and Troplong-Mondot in St-Émilion, as well as wine merchants. Both of those estates are now in other hands, and Thierry moved to Castillon in 2000, converting to first organics and then biodynamics from 2001. He is now certified with both Biodyvin and Demeter.

He joined the family business as a fourth-generation winemaker only after trying his hand as a jazz dancer and musician; he still plays a big part in organizing local music festivals. And I guess it is this slight outsider status that gave him the determination to follow a farming method that takes imagination and sensitivity. Today he notes the upsurge of interest towards greener winemaking even in Bordeaux. He once told me that after ten years of biodynamic farming in Castillon, he had never had the slightest interest or enquiry from any prominent producers as to its benefits, but that all of this has changed over the past few years.

'There is an undeniable groundswell towards biodynamics now.'
Thierry Valette

CLOS PUY ARNAUD CASTILLON-CÔTES DE BORDEAUX 2011, FRANCE ✱ ✱

There are some that I feel the same way about as a Haruki Murakami novel. A little proprietorial to be honest, and always a little worried to open a new bottle in case, this time, it disappoints. Clos Puy Arnaud certainly doesn't here, even in the 2011 vintage (often overlooked in Bordeaux), which is tasting delicious right now. 70% Merlot, 30% Cabernet Franc; rich, silky, end-of-summer fruits delivered with purity and concentration. This hums with energy and yet still delivers the power and sexiness you would expect from a Castillon wine. Brilliant match to turmeric chicken.

DOMAINE ALAIN GRAILLOT CROZES-HERMITAGE 2014, FRANCE

WWW.DOMAINEGRAILLOT.COM

Alain had a career as an engineer in the agro-chemical industry before his conversion to organic winemaking, so he knows what he's talking about. Today he works entirely organically alongside his son Maxime. The wines can take a good ten years of bottle age, so if you drink them young (like this one), get them into a carafe for a few hours first. Beautifully balanced Syrah, smoky and rich, with olives, blackberries and rosemary, aged in barrels but with just 10% new oak. Look out also for his Crozes-Hermitage La Guiraude, a limited-production bottling that is a little more tight and structured, with liquorice concentration in spades. Match with chicken and green olives, served with rice.

DOMAINE CLOS ROUGEARD LE BOURG SAUMUR-CHAMPIGNY 2010, FRANCE

15 RUE DE L'EGLISE, 49400 CHACÉ

This Loire estate manages to pull off the trick of being completely under the radar and yet one of the most sought-after cult wines in France (and beyond). No surprise really, because if certain Burgundy estates can claim to reveal the true heart of Pinot Noir, then these guys have every right to make the same claim for Cabernet Franc. Hard to say exactly what makes this wine so different. The texture perhaps, which is silky without being polished, yielding without being soft. It's fleshy, rich in berry fruits yet subtle as you like. The brilliant co-owner Charly Foucault died in 2016, after running things with his brother Nady since the 70s, and it has now been confirmed that Clos Rougeard has been bought by a rather more high-profile set of brothers, Martin and Olivier Bouygues. Let's hope they continue to show us just how incredible a wine this can be. No chemicals have ever been used on the soils here. Bring out the big guns of lamb or steak as a food match, or just savour with duck rillettes and crusty bread.

DOMAINE COMBIER CROZES-HERMITAGE 2014, FRANCE

WWW.DOMAINE-COMBIER.COM

One of the oldest organic estates in France; celebrating almost 50 years of farming without chemicals coming anywhere near the land. It has become a symbol of the renaissance of Crozes-Hermitage, where today around 30% of the wine is grown from organic grapes. This is the estate's classic 100% Syrah, showing plenty of Northern Rhône character while retaining an easy-to-approach drinkability. Juicy, velvety autumnal fruits merge beautifully with a palate dominated by floral tones and earthy spice. A perfect accompaniment to sticky pulled pork or caramelized oven-roasted vegetables.

DOMAINE DE BELLIVÈRE LE ROUGE-GORGE COTEAUX DU LOIR 2015, FRANCE

WWW.BELLIVIERE.COM

Made from the Pineau d'Aunis grape, with vines aged between 25 to 40 years old, the word that I hear most from friends when tasting this wine is 'authentic'. I think what they mean is that it's one of those bottles that transports you right to the fields where the vines grew. Christine and Eric Nicolas certainly spend most of their time out in the vineyards, which they created from scratch in 1995. The soils for this bottle are clay and flint on limestone (that specific Loire version known as tuffeau), worked to ensure low yields. It majors on an almost sour-cherry expression that takes a minute to get used to but then gives this incredible mouthwatering flourish, plush with spicy red fruits and bracken.

'Pair this with an Osso-Iraty, a sheep cheese from the Pays Basque served with black cherry confiture, or a sashimi of tuna.'
Pascaline Lepeltier MS, Rouge Tomate, New York

DOMAINE DE LA POUSSE D'OR CHAMBOLLE-MUSIGNY PREMIER CRU LES AMOREUSES 2014, FRANCE * * *

WWW.EN.LAPOUSSEDOR.FR

Patrick Landanger has owned this legendary Burgundy property since 1997, making *premier* and *grand cru* bottlings from a range of giddyingly lovely spots. This tiny plot of less than 0.2 ha of Pinot Noir vines is the oldest planted in 1976. Aged in 30%-new-oak barrels for 15 months and bottled both unfiltered and unfined; this 2014 is of course still a baby, especially as this particular bottling is known to take its own sweet time in ageing. But already you can feel the layers of fruit, bursting with tight berry flavours – absolutely gorgeous wine that you feel so lucky to drink. This is Pinot Noir at its most textured and lush, with earthy touches. When thinking about pairing, Pinot is an extremely flexible food wine, because it is supple, fruit-filled and rarely weighed down with heavy oak flavours. It works with fish as well as meat, so this is good to remember if you are splitting a bottle at a restaurant and ordering different dishes. This bottle is complex and nuanced, and just gorgeous with a lobster ravioli, a delicate plate of confit duck or even a simple caramelized onion tart. The estate is currently in the process of certification with Demeter, which should be complete in 2018, but it has been practicing organics and biodynamics for 20 years.

Aubert de Villaine
Domaine de la Romanée-Conti

1 PLACE DE L'ÉGLISE, 21700 VOSNE-ROMANÉE, FRANCE
WWW.ROMANEE-CONTI.FR

It tells you something about Aubert de Villaine when he says that 50% of his heart lies in the Côte Chalonnaise. This is a man who truly merits the term wine legend, owner of Domaine de la Romanée-Conti in Vosne Romanée, and yet who is only too thrilled to talk about the more discreet estate that he started with his American wife down in the less fashionable southern part of Burgundy. 'Bouzeron is a more modest appellation,' he says, 'but that means nothing. The winemakers there have just as much heart as the rest of Burgundy.'

This is utterly expected from a man who champions the history and potential of Burgundy in all its forms. Softly spoken with a regal bearing, his attitude towards biodynamic winemaking in his own properties is only natural for someone who sees not only the big picture of Burgundy but also the tiny details that make up old-school artisan farming. Aubert began experimenting with biodynamics back in the 80s, after becoming organic in 1985.

The estate has been 100% biodynamic since 2006, but was only certified in 2017. 'Friends eventually convinced me that being certified was worth it,' he says with a smile.

'At first I was interested only in the benefits that it brought to the terroir, but the certification perhaps underlines the importance of being totally committed to it.'
Aubert de Villaine

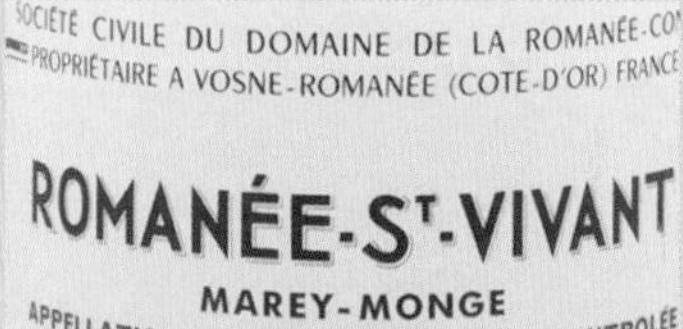

DOMAINE DE LA ROMANÉE-CONTI, ROMANÉE-ST-VIVANT 2009, FRANCE * * *

For many years Aubert de Villaine resisted getting certified at Romanée Conti (aka DRC), but, as of 2017, he is officially recognized by Demeter. Romanée-St-Vivant is one of the six *grand-cru* reds that Aubert makes from his Vosne-Romanée property, sharing this *grand cru* between nine other owners. Aubert has the largest slice of the pie at 5.28 ha. These are special occasion wines, it goes without saying, but serve as proof that wine, at its best, can provide memories that last in ways that few other drinks can match. The aromatic complexity curling out of this particular glass is incredible: soft, dense, floral, seductive, refined, violets, blackberries, green tea, woodsmoke. It deserves at least another ten years in bottle before it really starts to open up. But, whenever you drink it, expect elegance over finesse, complexity over ease, and expect to be smiling for days. To be honest, I spent weeks trying to decide what food would be best to pair with this, and in the end went with roasted wild mushrooms with a handful of fresh herbs in a risotto. Grouse, partridge or any game dishes would work amazingly well too.

Thierry Germain

Domaine des Roches Neuves

56 BOULEVARD SAINT-VINCENT, 49400 VARRAINS, FRANCE
WWW.ROCHESNEUVES.COM

Thierry Germain is a winemaker's winemaker – even more so because he, as his friends like to tease him, had to leave his home town of Bordeaux to learn about terroir in the Loire. That was back in 1991 when he was 24 years old, son of a fifth generation winemaker on the Right Bank of Bordeaux. Today it's impossible to think of him anywhere else but in the pretty village of Varrains, where his vineyard has slowly and steadily become a polyculture farm, inspired by his belief in a balanced ecosystem, and the influence of his spiritual father, the late Charly Foucault of nearby Clos Rougeard.

His two stocky shire horses plough the vines to lessen any soil compaction, and are joined by specially designed quad bikes that can move even more lightly across the soil than horses hooves. Clearly Thierry is not someone who stops at the first answer, or is afraid of a little challenge (he dug out the cellar below his winery, for example, pretty much by himself). This is why he has made such an impact on his adopted region in such a (relatively speaking, this is winemaking after all) short space of time. Listening, observing, experimenting, always in search of wines with as little between them and the land that grows them as possible.

'We have a habit of hiding our faults. It's human nature. And it's also true in winemaking. Instead I try as much as I can to minimize interventions that will cover up a wine's true nature.'
Thierry Germain

DOMAINE DE ROCHES NEUVES FRANC DE PIED SAUMUR-CHAMPIGNY 2015, FRANCE

Way before I knew about biodynamic wines, I knew Thierry Germain's bottles tasted different: brighter, fresher, more vibrant than average. No surprise really, as he is one of the leading biodynamic producers in France, and it's almost impossible to go wrong when choosing his wines. This is an unusual bottling that comes from ungrafted Cabernet Franc vines grown in a horse-ploughed field. The hand-harvested fruit is whole-bunch fermented at low temperatures to emphasize the pure-fruit expression – said to result from growing vines on their original rootstock. The results give beautifully clean and bright redcurrant and violet flavours and it makes a great match for Chinese crispy duck.

DOMAINE DE TRÉVALLON IGP ALPILLES 2013, FRANCE

WWW.DOMAINEGRAILLOT.COM

A Provence legend owned by Eloi Dürrbach, made from an unusual blend of equal parts Cabernet Sauvignon and Syrah. This was virgin land ('the kingdom of *garrigue*' as they poetically put it) before Eloi's father René Dürrbach (a renowned painter and sculptor) and his wife Jacqueline bought first Mas Chabert and then the neighbouring property Trévallon in 1960. In 1973 their son Eloi left Paris to follow his heart to this remote corner of Provence, and basically single-handedly created a vineyard by clearing the rocky slopes of the Alpilles. The first vintage was 1976, and ever since he's been steadily doing things his own way, crafting one of the great (relatively) unsung wines of France. Today Trévallon covers 15 ha of red grapes and 2 ha of white, with the next generation Antoine and Ostiane part of the team. This 2013 has elegant tannins, soft but hugely expressive black-berry fruits, and is layoured with liquorice, figs, Mediterranean scrub and black pepper. It is released after two years of ageing and is bottled unfiltered. Match with a slow-cooked lamb stew perhaps.

Above: Domaine de Trévallon is found on the northern side of the Alpilles mountain range in Provence.

Julien Guillot

Domaine des Vignes du Mayn

RUE DES MOINES, SAGY-LE-HAUT, 71260 CRUZILLE, FRANCE
WWW.VIGNES-DU-MAYNES.COM

You know someone is a serious winemaker when they tell you they were born 'at harvest time forty-two years ago', and that their earliest memories were doing harvest with their father. But Julien Guillot is also something of a surprise. Before joining the family business of wine, he had a brief career as an actor, first as a child in a travelling theatre troupe and then as an adult – Molière, Caberet, stand-up comedian, acrobat. His sister still works on the stage.

'By nineteen I knew I wanted to come back to the family estate, and for a while I did both, until at the age of twenty-eight I returned full-time to Burgundy. My father was right at the origin of the organics movement in France, and I remember (the oenologist and father of natural wines) Max Léglise being a regular visitor when I was young. Both of their ideas were always a part of my life, and even back then they were working without added sulphur.'

'I don't want to be boxed in to the term natural winemaker, for me it is just the way to make wine. From the theatre you learn how to put yourself in the skin of others, to have compassion and try to understand something beyond yourself, and these are lessons that I still try to use. Being an organic winemaker is also about being in tune with something beyond yourself.'

> 'This terroir has always been protected from chemicals, and I really try to make the wines that come from it as authentic as possible.'
> *Julien Guillot*

DOMAINE DES VIGNES DU MAYNES CLOS DES VIGNES DU MAYNES CUVÉE AUGUSTE 2014, FRANCE

The brilliant and talented Julien Guillot is proof that in Burgundy you don't always have to go to the most expensive and renowned appellations to get truly amazing wines, full of rich fruits. This wine ripples with sexy, earthy flavours – its Pinot vines are the result of massal selection dating back to 1964. Organic since 1954, biodynamic since 1998, Julien can also slip into the natural banner as he uses indigenous yeasts, no added sulphur and minimum intervention at every step. Based in the Mâconnais, his father Alain was president of the National Federation of Organic Wines, which helped convince France to usher in the first organic certification programme in the 80s. Let's just say you're in safe hands. I drank it with grilled chicken and taboulé, but it could easily have stood up to richer flavours.

Franck and Frédérick Buisson

Domaine Henri et Gilles Buisson

IMPASSE DU CLOU, 21190 SAINT-ROMAIN, FRANCE
WWW.DOMAINE-BUISSON.COM

Franck and Frédérick are the next generation to take over here, although their father Gilles still lives on the property and helps out even in his retirement. Franck has a Master of Commerce and Frédérick a Master in Oenology, but it's pretty much all hands on deck both in the vines and in the cellar.

'The move towards organic was already important for our grandfather Henri,' says Franck. 'He was working with François Bouchet and rejecting the idea of chemicals. Then in the eighties our father took over and the trend was for modernity, so the estate moved to conventional farming, until Frédérick joined in 2000. He had been working with Anne-Claude Leflaive for a while, and took the idea of organics back to our father. We tried our first bottling without sulphur in 2004, more out of intellectual curiosity.'

'There is a real sense of achievement to make a stable wine in this way, but we apply the same philosophy of minimum intervention and capturing the spirit of the fruit in all our wines.'
Franck Buisson

DOMAINE HENRI ET GILLES BUISSON SAINT-ROMAIN ABSOLU 2015, FRANCE * *

This, the eighth generation, is making some of the unfussiest, most deliciously drinkable wines you can imagine. The vines are located towards the south of Burgundy, in a fairly cool spot at an elevation of around 150m (492ft) higher than the rest of the Côte de Beaune. This has beautiful integrity to the fresh red fruits, peppered with spicy cloves and gingerbread. Vibrant and alive, you really feel it breathing through your palate. The estate is organic across all its wines, but only makes two cuvées with no added sulphur (around 5% of their entire production). 'We tried it out of curiosity,' says Franck Buisson, 'and loved the results'. This is a great wine, savoury and sappy and made for something simple and juicy like a grilled chicken and aubergine salad.

Opposite: Domaine Henri et Gilles Buisson is making fantastic wine out of Saint-Romain in Burgundy.

DOMAINE GAUBY MUNTADA IGP CÔTES CATALANES 2014, FRANCE

WWW.DOMAINEGAUBY.FR

Gérard Gauby created this estate in the 80s, and slowly but surely it has become one of the most exciting properties in southern France. This wine is subtle, fragrant, sculpted, incredibly balanced and easy to drink; with touches of rusticity that give charm and the wild abandon of great Syrah that reaches out for you. Give it time to open, but it builds over time, and will reward patience with peonies, fennel, liquorice (all flavours to pick up with your food choice). Half of this large 90 ha estate is planted to vines and the rest left wild – *garrigue* so typical of the Meditternean – close your eyes and you can smell it in the glass.

DOMAINE JACQUES PRIEUR VOLNAY CLOS DES SANTENOTS MONOPOLE PREMIER CRU VOLNAY 2014, FRANCE

WWW.PRIEUR.COM/EN

A winemaker to watch, Edouard Labruyère has been quietly making inroads into some of the most closed-off regions in France. His family started in Moulin-à-Vent, and now run estates in Pomerol, Burgundy and Champagne. This one, and Chateau Rouget in Pomerol, have been run entirely organically since Edouard took over, with a large slice of biodynamic practices. In 2016 he had to treat to avoid losing the entire harvest, but it is back to organics today. The 2014 vintage had hail problems (so small volumes); it is one of the earlier drinking vintages from this property. Expect fresh but intense redcurrant and raspberry fruit, liquorice root, smoky violets, fine tannins, with a focus on teasing out the fruit essence. The sweetness of the fruit would work great with a goat's cheese salad.

DOMAINE LOUIS CLAUDE DESVIGNES JAVERNIÈRES MORGON 2015, FRANCE

WWW.LOUIS-CLAUDE-DESVIGNES.COM

Brother and sister Claude-Emmanuelle and Louis-Claude Desvignes represent the eighth generation to make wine here. Javernières is from a plot at the base of the Côte de Py (essentially an extinct volcano), so still with plenty of granite and decomposed schist but also a little more clay and a large dose of iron oxide. Combine this with their careful, non-interventionist winemaking and you get both power and a beautiful floral character. Harvested late in the season, 100% Gamay has a fairly long fermentation with (almost entirely) whole bunches. The pure expression of raspberry and black-cherry fruit is given layers of complexity by a smoky caramel edge. I had this with a vegetable tagine, served with almonds, chickpeas and some wonderfully sweet dates.

DOMAINE MICHEL LAFARGE CLOS DU CHÂTEAU DES DUCS PREMIER CRU VOLNAY 2013, FRANCE * *

WWW.DOMAINELAFARGE.COM

Frédéric and Chantal Lafarge are now running this legendary Volnay property, having been handed on the pleasurable duty from Frédéric's father Michel. The vines are from some of the best sites of Volnay, worked biodynamically since 1990, Demeter-certified since 2000. The fragrance lifts you off your feet, and complex, wistful flavours are of late-summer fruits layered with softly smoked liquorice root. Light fining, no filtration, almost no new oak (5%). This wine is a monopole, in the family for just over 100 years.

'Elk carpaccio with lingonberries, Västerbotten cheese, pickled chanterelles and juniper cream.'
Fredrik Lindors, Sommelier, Grand Hotel, Stockholm

DOMAINE RICHAUD TERRE DE GALETS CÔTES DU RHÔNE 2014, FRANCE *

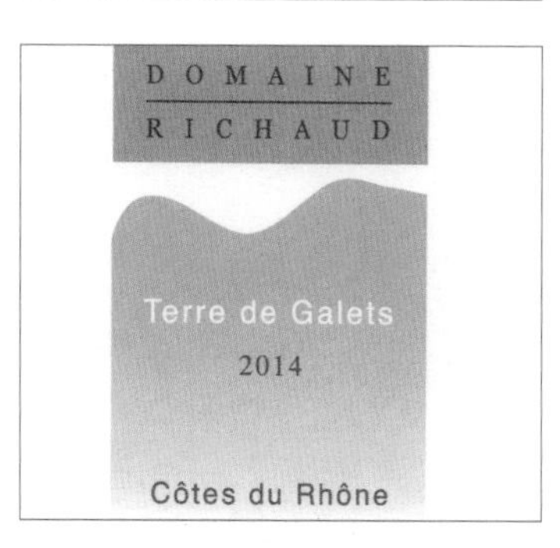

470 ROUTE DE VAISON LA ROMAINE, 84290 CAIRANNE

This is just utterly delicious, easy to drink, a great-value Rhône. You get the spice, the dirt, the juice-filled red fruits, the herbs and the grip of a Rhône Syrah, and at brilliant price. A generous, gorgeous wine. Unfiltered, unfined, tiny quantities of sulphur (usually under 15ppm). No need to wait to drink it either, it's perfect now, and will be hugely flexible with food, from burgers to pizza to a slow-roasted rib of pork.

DOMAINE TEMPIER CUVÉE CLASSIQUE BANDOL 2015, FRANCE

WWW DOMAINETEMPIER.COM/EN

This was a hot vintage, yet Domaine Tempier managed to craft a beautifully balanced, elegant wine that showcases just how nuanced and enjoyable the wines from this Bandol powerhouse are. Long-term winemaker Daniel Ravier says it was the influence of Mediterranean breezes, helped by the soils that were still holding water reserves from 2014. The Cuvée Classique is a blend of 75% Mouvèdre with Grenache, Cinsault and a touch of old-vine Carignan. Full of the grace of Mourvèdre, the tannins are elegant and well-drawn; fresh and sappy it is totally moreish with wild cherry and a brush of Mediterranean herbs. This shows exactly why some of the best winemakers in France have a few cases of Domaine Tempier stacked away in their cellar – it's a winemaker's wine. I matched it with a wild boar ravioli recently, but have enjoyed various vintages with a whole variety of foods over the years – one memorable pairing being pulled pork cooked for almost 12 hours on the barbecue.

Mathieu and Camille Lapierre
Domaine des Chênes

69910 VILLIÉ-MORGON, FRANCE
WWW.MARCEL-LAPIERRE.COM

A profile of Mathieu and Camille has to include a few words on their father Marcel, who took over his family estate in the 70s, but changed it profoundly in 1981. This was when he met Jules Chauvet, a winemaker, chemist and researcher who became known for his fight against the use of chemical fertilizers. Marcel became one of the Club of Four, a group of Beaujolais winemakers (with Guy Breton, Jean-Paul Thévenet and Jean Foillard) who stood by the principles of traditional viticulture – old vines, no herbicides or pesticides, minimal sulphur and no additions of any kind such as chaptilization or acidification.

Marcel passed away in 2010, but his children Mathieu and Camille are continuing his work. Mathieu trained as a chef but worked alongside his father from 2005, with Camille joining in 2014 after first working as a sommelier and then finishing her oenology degree (they have another sister Anne who is equally crazy about wine but who is not yet at their sides). It was this young generation that decided to get organic certification and to begin working biodynamically, and who continue to shape the future of these wines that are a profound signpost to why Beaujolais should be taken seriously. And to uphold an approach that their father liked to call 'soft winemaking'.

ALC.13,5% BY VOL. 2016
Morgon
M.&C. Lapierre à Villié-Morgon (Rhône)

MARCEL LAPIERRE MORGON 2016, FRANCE

Marcel Lapierre was something of a hero to the natural-wine movement, and his Morgon has been called 'The Original Natural Wine'. Now run by his children Mathieu and Camille, this delivers star-bright, sappy, mouthwatering dark fruit. It has an irresistible energy, with a clear sense of minerality. It is 100% Gamay grown on granite soil, picked late in the season to ensure full ripeness and aged on the lees for several months in large oak casks with zero additives at any point. Joyful wines and the perfect accompaniment to a lazy lunch of salads and quiche, a large table and lots of friends.

M&S BOUCHET FLEUR BLEUE VIN DE FRANCE 2015, FRANCE

RUETTE DU MOULIN, 49260 MONTREUIL-BELLAY

If you ever have to pick a winemaker's winemaker in the biodynamic world, you've hit pay dirt here. Mathieu is the son of François Bouchet, one of the towering figures of biodynamics in 20th century France, a disciple of Steiner who has been a consultant to many of the estates in these pages. So it's not surprising that the family estate has been farming biodynamically since 1962. The three wines of the property are made from old vines aged between 60 and 100 years old; aged in large-sized old oak barrels in caves of tuffeau, the classic limestone of the Loire. Nothing is racked, fined or filtered and any sulphur use is strictly minimal. All of this means that they are not always popular with their local appellation controlée in Saumur, so since 2008 they have bottled everything as Vin de France. Their Blanc de Chenin comes from 100-year-old vines and is utterly incredible, as is the Sylph Cabernet Franc, but I wanted to draw your attention to this unusual wine from the Grolleau grape, which is more usually used to make rosé. Mathieu has teased out real depth and flavour from these 80-year-old vines – a little wild, certainly, but firm and fresh with soft blueberry and cherry blossom shot through with gentle mushroom and undergrowth notes. Rustic in all the right ways, it worked perfectly with a grilled root vegetable and goat's cheese salad.

NO CONTROL VINCE VIGNERON LA COULÉE VIN DE FRANCE 2015, FRANCE

WWW.LOUIS-CLAUDE-DESVIGNES.COM

This is a sommeliers' darling and tastes like it, packed full of crunchy fruit that transports you to a smoky Parisian bistro or a Williamsburg wine bar – and right on cue winemaker Vincent Marie describes it as, 'natural wine, free of chemical products, only grapes, passion and rock n' roll'. But one taste and you totally get it, this wine has the iron-rich, herb-strewn taste of the wild side of Syrah, here mixed with perfectly juicy Gamay. All from vines in the Puy-de-Dôme region of central France, which have been farmed organically for over 25 years (only 'No Control' since 2014, though some of the vines are well over 100 years old), this is rather brilliant, a wine that you keep going back to for more. There is no added sulphur here. Food-wise, think of concentrated but clean flavours – swordfish steak, grilled sardines, anchovy toast. Vincent was one of the organizers of a natural wine fair in Caen, on the Atlantic coast of France, and eventually got such a bug for it that he decided to make his own wines. He headed off to Alsace, where he studied winemaking but always stuck to the idea of making natural wines. He now puts his theories into practice in these bottles.

Stéphane Ogier

Domaine Michel et Stéphane Ogier

3 CHEMIN DU BAC, 69420 AMPUIS, FRANCE
WWW.STEPHANEOGIER.FR

Stéphane Ogier is a winemaker that I always look forward to seeing again – whether in person or through his wines. Since coming back to Côte-Rôtie in 1997, fresh from studying and then winemaking in Burgundy and South Africa, he has been slowly but surely helping his father Michel to turn Domaine Michel et Stéphane Ogier into one of the highest quality and most exciting properties in the Northern Rhone.

Burgundy has influenced Stéphane in many ways. Not only did he study winemaking in Beaune, but he vinifies his dozens of different plots separately, according to their exposition and soil type. The two signature names of the Ogier property are the iconic vines of Lancement (Côte Blonde) and La Belle Hélène (Côte Brun) – 'the two great, opposing terroirs of Côte-Rôtie', as he puts it – but the signature of the property is their estate Reserve blend.

Walking, or scrambling, across these steep slopes with Stéphane is a lesson in artisan production. His plots are mainly orientated east, and many were replanted by Stéphane himself, with help from a stone mason who constructed or restored stone walls that protect the steep slopes from erosion.

'I don't have any organic label certification, but I work in the philosophy. I never use anti-rot or insecticide treatments, I always use organic fertilizer and all the tilling of the soils has been done either by horses or by hand (*à la pioche*) for the past seven years.'
Stéphane Ogier

STÉPHANE OGIER L'ÂME SOEUR IGP COLLINES RHODANIENNES 2013, FRANCE ✱

If you like Côte-Rôtie Syrah, try the wines from Seyssuel on the other side of the Rhône River, sharing much of the same elevation, orientation and soil type as their more famous neighbours. Stéphane Ogier is one of the most brilliant winemakers in France, and though he is a little more expensive than most other Seyssuel producers, this wine is still great value when you look at the prices of his celebrated Côte-Rôties. He doesn't have any organic certification, but he farms with no chemicals in the vineyard, everything is done by hand, any fertilizers are organic, and all the soils are worked by hand and have been since he joined his father at the property over seven years ago. This is just heavenly Syrah: fragrant violet and sour cherry, textured, structured without being intrusive; it rises vertically through your palate, caressing it along the way, and doesn't let go. Try herb-encrusted lamb cutlets, or a Chinese twice-cooked pork could also work.

Opposite/Overleaf: Stéphane Ogier is a name to know in the Northern Rhône.

RENAULT

Hansjörg Rebholz
Weingut Ökonomierat Rebholz

WEINSTRASSE 54, 76833 SIEBELDINGEN, GERMANY
WWW.OEKONOMIERAT-REBHOLZ

'We started working organically because it was the only logical way to follow my grandfather's idea of natural wines,' says Hansjörg Rebholz. 'His idea was to influence the wine as little as possible.'

Hansjörg is the third generation of his family to work the Pfalz vineyard, and has been named Winemaker of the Year several times in Germany for his careful, precise and unfussy commitment to quality. His grandfather Eduard Rebholz was a scientist who researched climate, soils and vinification methods, but who spoke out against typical German practices at the time such as the adding of *süsreserve* (unfermented grape juice) to sweeten the wines.

'The most challenging thing about organics is that it doesn't work in the same way for all vineyard sites,' Hansjörg says. 'It takes time to get a know a vineyard and be able to "feel" what it needs. At the same time global warming is changing the types of pests that we need to deal with, using only organic treatments can be tough. But the positive points make it worthwhile. The wines are much higher in minerality, have more length and ripe, harmonious acidity with the right amount of salinity. We really feel the impact of the vineyard site in the wine, and it is easier to reach the point of perfect ripeness with less sugar content, so the wines are lighter in alcohol and more elegant.'

'Every year we are enthusiastic about the diversity of animals and insects in the vineyards. Each terroir has its own distinct native floral and fauna, and that's exactly what we're after.'
Hansjörg Rebholz

WEINGUT ÖKONOMIERAT REBHOLZ
SPÄTBURGUNDER TRADITION TROCKEN
PFALZ 2014, GERMANY

Lovely sweet and juicy Pinot Noir fruit from Hansjörg and Birgit Rebholz. A soft cherry red, it delivers delicate tannins and an elegant fruit structure – such a silky texture you can totally lean into this wine. This red wine is totally flexible with meat, fish or vegetable dishes. I had it at lunch with a delicate spinach, goat's cheese and pomegrante salad, and it was utterly delicious. Hansjorg has been a serious voice in the move towards quality dry Rieslings, but here he shows his light touch with Pinot.

Opposite: Weingut Ökonomierat Rebholz focuses on making wines with as little intervention as possible.

COS PITHOS ROSSO 2014, ITALY

WWW.COSVITTORIA.IT

Pithos is the local Sicilian name for amphora, and COS winery works with the best, as theirs are made by Juan Padilla, a craftsman located near to Madrid who supplies top producers world over with these increasingly sought-after clay pots. The wine itself is a blend of 60% Nero d'Avola and 40% Frappato Victory, so one tannic grape against one far more floral and feminine style, aged in 500-litre amphoras that are buried in the ground to keep the temperature cool. The two grapes really do play sensually off each other – with fragrant Morello cherries against a powerful smokiness. A beautiful touch of natural bitterness on the finish is emphasized by the vibrant acidity, This is meaty at first, masculine and powerful, but the floral side comes out after ten minutes in the glass. Makes you think, and makes you reach for another sip. Grilled tuna or even sardines would be a good match.

FÈLSINA BERARDENGA DOCG CHIANTI CLASSICO 2011, ITALY

WWW.FELSINA.IT

I first tracked this down after a recommendation from Eric Asimov in the *New York Times*, one of my favourite wine writers. He said it was one of his go-to staples for everyday bottles, which was good enough for me. This is stuffed full of wild-berry fruit, soft violets and gentle acidity at an extremely reasonable price. It is one of those wines that tastes effortless and unforced, with silky tannins and mouthwatering fruits. The vineyards are located in the southeast of the appellation, to the southeast of Siena, between 320 and 420m (1050 and 1378ft) above sea level. 100% Sangiovese, fermented in stainless steel and aged in Slovenian oak barrels for 12 months. Grilled red peppers and chorizo soaked in olive oil would be irresistible with this wine.

Elisabetta Foradori
Foradori

VIA DAMIANO CHIESA 1, 38017 MEZZOLOMBARDO, ITALY
WWW.ELISABETTAFORADORI.COM

I have to admit that I had only heard about Elisabetta Foradori by reputation before getting to know her wines for this book. They have been a revelation, and her low-key, steadfast approach to winemaking totally inspiring. The winery itself is set in the Dolomite mountains of northern Italy – craggy granite outcrops are everywhere around as you look from the vines – and dates back to 1901. It has been in the Foradori family since 1929, with Elisabetta joining after finishing winemaking studies in 1984.

She is happy to share that when she first came home, she was focused on the winemaking and not on the vineyard, 'but at a certain point I started to realize that I didn't get excited by my own wines because they lacked soul. I am so grateful now to look back at where I was, and how much it helped me to learn.'

She changed her focus to biodynamics in the 90s and ever since has practiced winemaking taken back to its essence. If you are lucky enough to visit, you are most likely to find Elisabetta outside in the vines, or tending to the trees and flowers that grow abundantly here.

'My focus today is leaving this land alive and healthy for my four children.'
Elisabetta Foradori

FORADORI SGARZON TEROLDEGO ALTO ALDIGE 2014, ITALY * * *

Elisabetta Foradori has quietly turned into one of Italy's winemaking superstars. You'll know why when you open a bottle: easy alcohol levels but a wave of graphite, wild cherries, grilled herbs, blueberries and a touch of rhubarb also thrown in. The fruit is juicy and full of integrity, just on the right side of fresh, with a touch of bitterness on the finish adds to the intriguing layers of complexity. I had this with homemade pasta and fresh tomato sauce, and was in heaven. Unfiltered, low sulphur, macerated with the Teroldego grape skins for eight months in clay *tinajas* then vinified with natural yeasts, from vines in the cool-climate Sgarzon vineyard. If you like this wine, I think we'll be friends.

Opposite: The hillside vineyards of Foradori.

Giuseppe Vajra

G.D. Vajra

VIA DELLE VIOLE, 25 - FRAZ. VERGNE 12060, BAROLO, ITALY
WWW.GDVAJRA.IT

It's hard to find a more likeable winemaker than Guiseppe Vajra. Who can't warm to a someone who says things like, 'all we ask for is ease, simplicity, and a wine that goes with food'?

> 'Good wine can sometimes be simple, sometimes complex, but it is always a pleasure.'
> *Giuseppe Vajra*

His fluency and warmth when talking about his family's wine is totally infectious in person, but I loved G.D. Vajra wines long before I went to visit them in their beautiful estate just off the main road that leads up to Barolo. (If you have a chance, go to eat at the little trattoria almost directly opposite their gates, where I had a totally unforgettable lunch of pasta with white truffles grated over it sat at a tiny outside table in October sunshine a few years ago.)

Guiseppe works alongside his father Aldo and various brothers and sisters. Their main vineyards are found in Vergne, one of the highest villages in Barolo, but they have plots elsewhere too, and are increasingly prominent in the move towards rescuing the forgotten grapes of Piedmont, and in reviving old techniques such as the Metodo Classico. They are even improving the quality of the more usual grapes such as Dolcetto and Nebbiolo through massal selection carried out on their estate, to encourage genetic diversity that will see the vines remain healthy for future generations.

G.D. VAJRA LANGHE FREISA KYÈ 2014, ITALY

If there is one name in Piedmont that I just keep going back to, it is G.D. Vajra. It became one of the first organically certified estate in Piedmont way back in 1971, and has consistently done its bit to protect the heritage of the region. This wine is proof of their effectiveness, and it is made from a local variety called Freisa, which was at risk of dying out (in the past it was used to make red Vermouth). These grapes were planted through massal selection by Aldo Vajra from a vineyard planted by his grandfather, now grown on the clay-rich San Ponzio vineyard, in the commune of Barolo at 390 to 410m (1280 to 1345ft) altitude. Such a discovery, it's a wine that I keep coming back to. Deceptively simple, it is deep-ruby red in colour, with an almost wild perfume of tobacco, roses and wild herbs. On the palate it's full of soft red berries, gentle tannins, with gentle acidity and white pepper spicing and hints of hay. It's small production, one to savour with fresh tomato pasta or a homemade pizza – keep things simple.

Opposite: The Vajra family vineyards have been certified organic since 1971.

Arianna Occhipinti
Occhipinti

SP68 VITTORIA-PEDALINO KM 3,3, 97019 VITTORIA,
SICILY, ITALY
WWW.AGRICOLAOCCHIPINTI.IT

'I love the natural part of making wine, how it makes me feel in contact with the land that I come from.'
Arianna Occhipinti

Still barely into her 30s, Arianna got started making wine at 21, beginning with just 1 ha of vines and then slowly but surely building up to the 30 ha of vineyards and olive groves she has today.

'I fell in love with the wine when I was sixteen, working with my uncle who has another winery, and then I started by myself. At the beginning it was just an experiment, for kidding around; but the results were good and I wanted to continue. From the beginning I wanted it to be natural, so farmed with organics. For me it is a great job, because I like to make things, to do things with my hands, and this is very practical. I can also learn about geography through wine, and the history of my area.'

OCCHIPINTI SP68 NERO D'AVOLA E FRAPPATO IGT TERRE SICILIANE 2015, ITALY

Arianna Occhipinti is making wines away from the famous Etna region of Sicily, in the more agricultural southeast of the island known as Vittoria. She began making wine in 2004, and has since won the Next in Wine prize awarded to young Italian winemakers under 35 years old. This wine gives you a clue as to why: lyrical and packed with personality; light, crunchy and supple with some beautiful bitter herbs and nettles on the finish. Made from the indigenous grapes Nero d'Avola and Frappato farmed biodynamically since 2009 and left with full leaf cover to protect from the hottest summer sunshine. 'It's stupid not to be organic in Sicily,' Arianna commented, 'we have the perfect climate.' She has shown similar practicality when naming this wine – the SP68 is the road that passes directly in front of her winery. I would pick a *spaghetti alla puttanesca* with this.

MONTEVERTINE PIAN DEL CIAMPOLO IGT ROSSO DI TOSCANA 2014, ITALY

WWW.MONTEVERTINE.IT

Made with mainly Sangioveto (Sangiovese with a local spin) grapes by Martino Manetti with consultant Paolo Salvi. Manetti's father Sergio was once part of Chianti Classico, but withdrew in the early 80s and set about ploughing his own furrow, as they say, with a range of brilliant wines that should be sought out at all costs. His son is now in charge, but the philosophy remains the same. This is the entry level bottling, a blend of 90% Sangiovese and 10% Canaiolo and Colorino, fermented in cement tanks then aged for 18 months in Slovenian oak barrels; unfiltered, unfined. Ripe redcurrant and blueberry fruits tease with white pepper spices and the Tuscan brush of slightly bitter herbs, all delivered with mouthwatering freshness. Just keep things simple here – a good dollop of a meat ragu over your favourite pasta.

QUERCIABELLA DOCG CHIANTI CLASSICO 2013, ITALY ✱ ✱

WWW.QUERCIABELLA.COM

This 100% Sangiovese is studded with beautiful herbal, scrubland notes and comes from producer Sebastiano Cossia Castiglioni, who has done brilliant work on producing site-specific Chiantis. It needs a little time to open, as the tannins are firm; this is classic, measured winemaking with stunning potential, firm grip and true Tuscan structure. It opens up to show savoury red fruits, floral edging and great authenticity – mouthwatering with a wild boar stew, or sausage and mash. Or perhaps an aubergine lasagne, as Sebastiano practices something called cruelty-free biodynamics, which bars the use of animal-derived products at any point from grape growing through to the cellar.

RIECINE DOCG CHIANTI CLASSICO 2014, ITALY

WWW.RIECINE.IT/EN

There's plenty of evidence that suggests wine was being made in the Riecine hills back in the 12th century or earlier, but its modern-day incarnation is thanks to Englishman John Dunkley and wife Palmina, who bought a run-down stone villa in 1971 with just 1.5 ha and began restoring it and replanting vines. Owned since 2011 by Lana Frank, the wines are made by Alessandro Campatelli and Carlo Ferrini. This is a light, sculpted style of Chianti, bursting with red fruits and the perfect amount of smoked herbs that bring focus to the finish of the wine and make it so utterly drinkable. I would go with nothing more than fresh pasta, olive oil, Parmesan and lots of freshly ground black pepper. Worth seeking out their IGT Toscana La Gioia also.

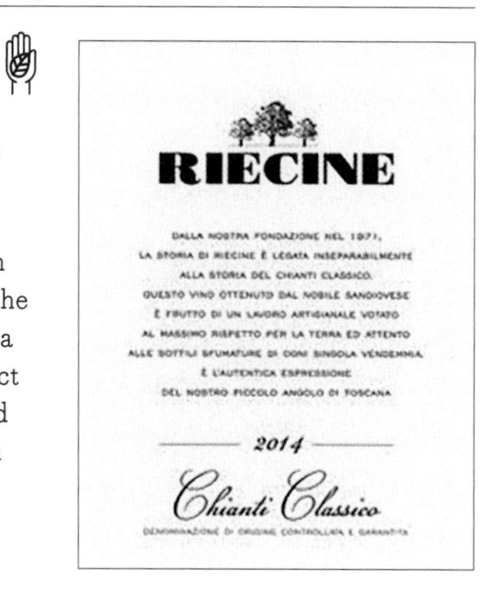

SERRAGGHIA ROSSO FANINO IGT TERRA SICILIANE 2012, ITALY

WWW.SERRAGGHIA.IT

Winemaker Gabrio Bini is doing pretty amazing things on the island of Pantelleria, off the southwest coast of Sicily and closer to Africa than any other part of Italy. As with Sicily, volcanic soils abound, and Gabrio works them entirely by hand, with help from the occasional horse. This wine is made from the local Catarratto and Pignatello grapes, fermented with their natural yeasts in clay amphora buried in the ground. This is not a wine that you simply take a sip of and put to one side – it pretty much demands you sit up and pay attention. It's refreshing in a way that you might not expect from an island so close to Sicily, but that's because this is a blend of 50/50 red and white grapes. What you get is a delicate, softly perfumed wine with fresh red fruits and savoury notes of oregano and rosemary that place it somewhere between a rosé and a light red. Take their advice for what to eat with this: capers with *fleur de sel* (they are grown on the island) and simple plates of salamis and olives.

SOTTIMANO BARBARESCO COTTÁ 2013, ITALY

WWW.SOTTIMANO.IT/EN

Wonderful wines from Andrea Sottimano. A relatively recent convert to a more hands-off approach to winemaking, today you find natural yeasts, 30% new oak for barrel ageing, unfiltered and unfined bottling – just pure unfettered fruit expression. The punch comes from the mid-palate and through to the finish, with a deepening of forest-floor, menthol and dancing-spice flavours. This vintage is restrained and balanced, but in most years the alcohol levels in the Cottá climb a little higher, even if they are never intrusive. I absolutely love this producer's Dolcetto d'Alba too if you are able to find it, usually extremely inexpensive, and a perfect fresh lunchtime bottle. Like all Piedmont wines, they are made for food – simple, fresh grilled vegetables would be perfect.

TENUTA LA NOVELLA CASA DI COLOMBO DOCG CHIANTI CLASSICO 2014, ITALY

WWW.TENUTALANOVELLA.COM

Organic for over 20 years, certified since 1998, this beautiful wine is a blend of 90% Sangiovese with just a 10% touch of Merlot, grown in an amphitheatre vineyard at 550m (1804ft) altitude. It is destemmed but not crushed for fermentation, and is aged for 12 months in French oak. Bright, fresh-berry fruit, with striking texture – it totally glides over the palate leaving persistent dots of cinnamon spice in its wake. Just add a plate of sliced prosciutto.

Opposite: The Chianti winery of Montevertine.

GUT OGGAU BERTHOLDI BURGENLAND 2013, AUSTRIA

WWW.GUTOGGAU.COM

These wines just seem to make everybody smile – not just the labels but the unforced character of the juice itself (a pretty much perfect reflection of winemaker Eduard Tscheppe). There is a great line-up of bottles to choose from, but I have picked the 100% Blaufränkisch as it's one of those Austrian grapes that is always on the edge of being better known but never quite getting there. Structured, complex, nicely balanced, revealing rich red fruits. A touch of salinity on the finish adds to the feeling of a supremely drinkable, food friendly wine. A plate of cured meats would do just nicely.

WEINGUT CLAUS PREISINGER PINOT NOIR 2015, AUSTRIA

WWW.CLAUSPREISINGER.AT

I have yet to open a bottle by Claus Preisigner that I haven't absolutely loved. This is extremely delicate, a little austere at first and then gently fragrant, clear redcurrant. It's a slow burn of a wine, step back and appreciate it, don't expect an immediate punch of flavour, but it's beautifully made, carefully placed and precise. Keep things simple, this is a great red wine for fish, but it has structure so stick to a sweeter, rich style like tuna in soy sauce, or a deliciously chargrilled mackerel.

WEINGUT JUDITH BECK PANNOBILE BURGENLAND 2013, AUSTRIA * *

WWW.WEINGUT-BECK.AT

Not content with going biodynamic in 2005, Judith Beck was the driving force behind the Pannobile group of nine Austrian winemakers who highlight traditional Austrian varieties from selected sites on the northeastern shore of Lake Neusiedl in Burgenland. Each winemaker bottles their own wine under their name; here you get 60% Zweigelt and 40% Blaufräkisch translated into a seriously grown up bottle. This is going to thrill you if you like a big impactful wine that still manages to deliver fine tannins and a whoosh of elegant freshness. Rich and spicy from the first nose, with some liquorice and minty touches, spicy olive paste, the whole thing is simultaneously intense and yet fresh and sculpted; 2013 remains extremely young and will age well. Carafe for a good few hours before drinking – and be totally relaxed about what to pair it with, as the fragrant and fresh body should work equally well with fish as meat.

INTELLEGO KOLBROEK, SWARTLAND SYRAH 2015, SOUTH AFRICA

WWW.INTELLEGOWINES.CO.ZA

These vineyards sit under an uncomfortably active volcano, which makes for extremely interesting soils if you can take the stress. And clearly Jurgen Gouws has no problem with that. A brilliant Syrah with an angularity and severity that is unusual to find in the New World. Your mouth literally contracts on the first sip, and then it opens up with a whoosh of fragrant peppery spice. It's fresh and focused while still being intricately flavoured and well paced. Food-wise, I'd grab something heavily spread in olive tapenade, anchovies, garlic, lemon, oregano… get those herb notes fully embraced.

Opposite, above: South Africa's Intellego vineyards sit on complex volcanic soils.

Lilian Carter
Wang Zhong Winery (Tiansai Vineyards)

WWW.TSJZ.COM

Named as one of the 2015 Future Leaders for Australia's wine industry, Lilian Carter splits her time between Melbourne and Xinjiang, on the edge of the Gobi desert. It takes more than 30 hours door to door, but then, as she has said on many occasions, she doesn't back away from a challenge. She grew up on a winery in Rutherglen, Victoria, but nearly became a landscape gardener before returning to her roots. Today she is winemaker at Wang Zhong Winery, owned by Chen Lizhong, where sunshine and dry conditions make organic farming achievable.

'But it gets cold in winter, so cold, right down to minus thirty degrees that you have to bury the vines to keep them alive until spring. That's the number one challenge of the area. However the extremely dry conditions are a friend to the organic farmer during the growing season. The irrigation water comes from the snow-melt and from underground and is delivered via drip.'
Lilian Carter

'The vineyard makes its own compost on a large scale, collected from local sheep, goats and pigs, mixed and managed in a purpose-built shed. They have invested a lot in the design and construction of a device that attaches to the tractor to rip into the ground and deliver the organic matter deep into the soil before planting, then as required into the mid-row.'

'There is a severe shortage of local labour and the added complication that workers and their families regularly migrate, following the harvest of different crops around the region. To overcome this challenge, accommodation and facilities has been developed to attract workers and to encourage them and their families to stay. This means that the significant investment in training workers in vineyard-management techniques like leaf plucking, shoot thinning and grape thinning is repaid year after year and the knowledge base is retained.'

TIANSAI VINEYARDS SKYLINE OF GOBI SELECTION MARSELAN 2014, CHINA

One of the very few organic wineries in China. Australian winemaker Lilian Carter is helping to shape the increasingly excellent output from this Gobi desert property (and if you think the desert heat is an issue, spare a thought for when a tornado ripped apart half of the vines last year). Tiansai produces an excellent Chardonnay that is worth looking out for, but my favourite in the range is the Marselan (a cross between Cabernet Sauvignon and Grenache). The wine has a lovely spiciness to it, with a medium body and light tannins that deliver intense flavour and great persistency without trying too hard to draw attention to itself. I would try an unusual pairing such as tea-smoked duck with this.

Opposite: Great wines in challenging conditions at Tiansai Vineyards in China.

天塞酒庄

Taras Ochota
Ochota Barrels

62 MERCHANTS ROAD, BASKET RANGE, SA 5318, AUSTRALIA
WWW.OCHOTABARRELS.COM

It is impossible not to be charmed by the aura of cultivated chaos that Taras of Ochota Barrels manages to get across. Here is a sample of one message I received while tracking down his wine:

> 'Someone hacked my email so I'm even more of a shemozzle than ever... plus I think it clever to reply to emails towards the end of an evening on the couch after wrestling with the joys of Calvados.'
> *Taras Ochota*

Put that together with his brilliant wines and subtly intuitive approach to winemaking, and it's hard to remain unmoved. Taras spent most of his teenage years playing bass in punk bands, before studying oenology at the University of Adelaide. He then worked in various wineries around Italy and California before heading home with his wife Amber (also a winemaker) to the Adelaide Hills.

Their philosophy, in their words, is to work organically farmed vineyards, 'planted to earth that is alive, emplying lo-fi techniques and picking decisions made purely on natural acidity.'

OCHOTA BARRELS IMPECCABLE DISORDER PINOT NOIR ADELAIDE HILLS 2015, AUSTRALIA

Delicate and fragrant, high acidity, lean savoury red fruit with touches of violets, this has a smoky, raw, earthy side (a friend I was tasting with called it 'feral') that is totally appealing and just a little bit confounding. The Pinot is grown on low-yielding old-bush vines 'picked basically when you can't stop eating them because they taste so delicious.' It's sappy, fresh and rather brilliant. Oh, and another beautiful wine in the range (they buy grapes as well as grow their own so there are usually a lot of different wines to choose from in any given vintage) is a Gamay called Price of Silence. Plenty of food options to choose from here, from grilled tuna to sweet and sour pork.

HENSCHKE LENSWOOD GILES PINOT NOIR ADELAIDE HILLS 2014

WWW.HENSCHKE.COM.AU

Prue and Stephen Henschke look after vines that are among the oldest still-producing in the world. Prue is easily among the most renowned viticulturists in Australia and studied botany and zoology at Adelaide University before spending two years in Germany at the Geisenheim Institute, renowned for its links with biodynamics. This 100% Pinot Noir is from grapes located in Lenswood, a cool climate site at 550m (1805ft) altitude and is aged in (30% new) French oak for ten months. 2014 was a tough vintage, but this wine is astonishing: holding itself back at first, asking you to stop what you are doing and pay attention. Slowly waves of redcurrants and dried roses head out, focused and delicate. Then juiciness, when black cherry and spice kick in. Keep things light to really focus on these flavours – a Spanish omelette and green salad would be about perfect.

PUNCH PINOT NOIR YARRA VALLEY 2013, AUSTRALIA

WWW.PUNCHED.COM.AU

These guys know all about climate change, as their winery was badly affected by the 2008 wildfires in Victoria. But with help from friends, the Lance family has got back to making the same brilliantly exciting wines that have been increasingly sought over for nearly 40 years. These dry-farmed Pinot Noir vines have a deep-root system and produce small amounts of intense, complex fruit. They go all out for natural winemaking and the result is a concentrated, caressing red-fruit character with a gourmet charred-oak edge adding layers of subtle complexity. This showcases the best of Pinot's lip-smacking moreish qualities and would be a great balance for simple but deep flavours like tuna sashimi. Not certified, but only organic-approved copper and sulphur sprays are used in the vineyard and no chemicals since 1997.

STEFANO LUBIANA WINES ESTATE PINOT NOIR TASMANIA 2015, AUSTRALIA ✱ ✱

WWW.SLW.COM.AU

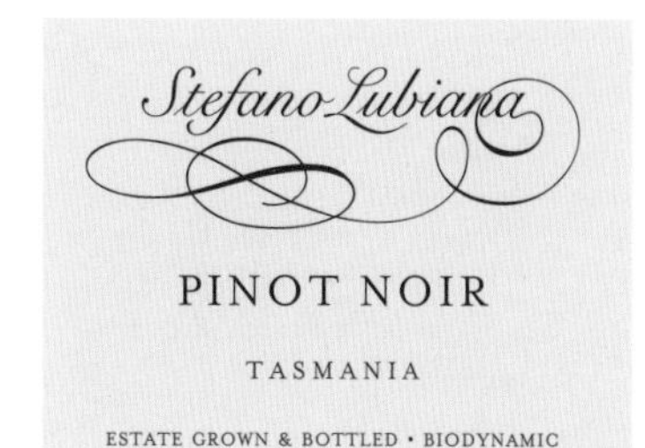

More proof of the potential of Tasmania as a wine-growing superpower. This is dripping with sweet Pinot fruit without being in any way over-done, it has the slightly wild cherry, earthy flavour of the Burgundian version of the grape, while still retaining the ripe character that comes from the southern hemisphere sunshine. The Lubiana winery is in the Derwent Valley near Hobart, with an Italian Osteria on site, serving free-range meat from local farmers, their own biodynamic vegetables and wild caught seafood from the waters of Tasmania.

TIMO MAYER YARRA VALLEY SYRAH 2015, AUSTRALIA

WWW.TIMOMAYER.COM.AU

Timo is a German winemaker who has been making wine in Australia for years. This wine has a smoky edge to the nose, and sings of the wild, blood-and-guts side of Syrah. A take-no-prisoners kind of a wine, gutsy, powerful, smokehouse-meets-herb-store, lavender and blackberries, not shy on the personality front. Wild yeast, open ferment, whole-bunch carbonic maceration, 16% new oak, with the rest of the barrels three years old, unfiltered and unfined. This will almost certainly split people into love it or hate it. Not a wine to ignore. It calls for crusty warm bread and a plate of chargrilled vegetables, drenched in olive oil and whatever spices you want to throw at it.

FELTON ROAD BLOCK 3 PINOT NOIR CENTRAL OTAGO 2014, NEW ZEALAND * * *

WWW.FELTONROAD.COM

One of the best-loved wineries in New Zealand, with reason. Block 3 refers to a specific plot of the Elms vineyard in Central Otago that has extremely complex glacial drift soils. The resulting wine is packed full of juicy, focused blueberry and redcurrant fruits, rich and yet earthy at the same time, with lovely touches of spiced chocolate. Sweeter and more succulent fruit than you would get in a Burgundy Pinot perhaps, but no less nuanced. Bottled unfined and unfiltered and yet delicate as you like, and with everything so integrated you can't tell where one flavour ends and another one starts.

'A very silky red wine but with enough weight and also an amazing array of aromas, especially cherry, wild berry and spices. Gorgeous with a tartare of tuna but only slightly spicy with a hint of soy sauce and a salad of avocado and tomato.'

Gerard Basset MW, Best Sommelier in the World 2010

SATO PINOT NOIR CENTRAL OTAGO 2014 NEW ZEALAND * * *

WWW.SATOWINES.COM

Yoshiaki and Kyoko Sato worked with leading biodynamic producers from Domaine Bizot in Vosne-Romanée, Pierre Frick in Alsace and Felton Road in New Zealand before setting up on their own in 2009. Yoshiaki likes wine with a light touch and moderate alcohol. This is tighter, leaner and more austere than many NZ Pinots, it needs time, but unquestionably this is great quality fruit, deeply finessed. Beautiful wine. You can feel the intellect and the searching here. A stunning match for crispy duck pancakes with plum sauce.

Overleaf: Burying cow horn manure for soil fertility (preparation BD500) is one of the signature biodynamic treatments, as here at Seresin Estate, New Zealand.

Clive Dougall
Seresin Estate

85 BEDFORD ROAD, RENWICK, MARLBOROUGH,
NEW ZEALAND
WWW.SERESIN.CO.NZ

Seresin winemaker Clive Dougall has a Maori tattoo on his right bicep and is winemaker at one of New Zealand's most iconic properties. So it's kind of surprising to learn that he grew up in Chiswick, London, and got the tattoo aged 18, 'because he liked the look of it'. Clive left school at 18 and ended up working in a wine shop, eventually heading his way up to managing the shop before deciding that he wanted to try his hand at making wine instead of selling it.

He finally made his way to New Zealand via Australia, and began to form his own views on winemaking, which, despite being a self-professed chemistry-geek, 'always seemed to me about intuition rather than exact science'. He arrived at Seresin in 2006 and has focused on making wines with purity and restraint ever since – now joined by his second-in-command Jordan Hoog.

'I love how every year the wines express something about the people and the places involved in that particular vintage. And how as a winemaker every day is different – one minute you're a chemist, the next an engineer, the next an artist and the next a farmer.'
Clive Dougall

SERESIN ESTATE RAUPO CREEK PINOT NOIR 2013, NEW ZEALAND

The winemakers at Seresin know what they are doing with Pinot; this is intense but pure and vibrant. Fairly high acidity, it partners rich food perfectly, where the freshness cuts right through (I chose homemade burgers, but a grilled steak would also work). Dark-berry fruits and plenty of spicy herbs, all supported by fine tannins; this is a lovely savoury wine. Owner Michael Seresin also has a successful career as cinematographer for a couple of films you might have heard of – *Bugsy Malone*, *Midnight Express*, *Gravity*, *Dawn of the Planet of the Apes*.

'Veal entrecote pan fried with garlic and butter, grilled courgette and glazed summer carrots and fennel, finished off with a flavourful sauce vierge made with a rich demi-glace/veal or beef stock and an intense (preferably organic) extra virgin olive oil.'
Henrik Dahl Jahnsen, Bølgen & Moi, Norway

California

04

Full & Warming Reds

Ronan Sayburn MS

Master Sommelier, 67 Pall Mall, London

I was born in Scarborough, North Yorkshire, where my father used to make wine out of parsnips, rhubarb, blackberries... whatever he could lay his hands on. Once a year he would bottle up the wine himself and store it in the garage, so you could say my interest in handmade, artisan wines started at a very early age.

I love that wine is a handcrafted product. Many of the most celebrated examples have been made by the same family for generations, from the same small patch of land. The inevitable respect for the soil that this results in seems reflected in all organic and biodynamic producers, whether they are seventh or first generation winemaker. It might be challenging to work this way, but the results are wines that represent the growers' dedication and commitment not just to their land, but to producing wines that put the focus on pure flavour and intensity of fruit.

These flavours deserve food to really bring them out. And the full-bodied tannic wines in this chapter generally work well when paired with rich proteins such as meat dishes and heavier sauces.

Extreme cooking techniques, such as flame-grilling or barbecuing, generally pair well with the most intense wines from warmer sunnier climates, where flavours and tannins are wrapped in ripe jammy fruits. Many New World wines (Australia, Argentina or California) and warm-climate European styles (Spain and southern Italy) fall into this category. They pair well with beef, venison or boar that is charred or pot-roasted with dark sauces that can be flavoured with dark berries, chocolate, juniper or liquorice. The ripeness of the fruit, the texture of the tannins and the level of alcohol is all important to consider.

Full-bodied wines from cooler environments (such as northern Europe or New Zealand) may be paired with less punchy robust flavours. These wines, while still having firm tannins and weighty mouthfeel, have less alcohol, less body and a leaner character. They often show more earthy notes and delicate fruit. Beef, duck, lamb and game are often good choices – pan-fried, slow-roasted, gently confit'd, encased in pastry or served with truffle and herbs.

The idea that pairing wines with dishes from the same country or region is very sensible as the styles of cuisine and styles of wine have probably grown in harmony together for many years.

The phrase 'it goes with where it grows' is apt when it comes to food and wine matching. For example, the salt-marsh lambs that graze on the banks of the Gironde estuary are perfect matches with wines from the Médoc; Piedmontese Nebbiolo from Barolo or Barbaresco are ideal choices with a hunter's catch of game birds, wild-boar stew and a garniture of Alba truffle. Sun-dried Italian tomatoes, flavoured with thyme and oregano served with pasta and meatballs are sublime with Chianti's Sangiovese grape.

The main event is really the meat or protein but it can take backseat at times depending on the intensity of the sauce. Echoing the flavours in wine can be successful; so making a sauce with a reduction of red wine or Port and flavouring with blackberries, bacon and black pepper for Syrah, blackcurrant and mushroom for Cabernet, cherry and cranberry for Tempranillo or Sangiovese can be a triumph. Australian Shiraz can be lifted with the addition of mint relish or by adding fresh rosemary to a sauce. Chilean Carmenère served with grilled beef and herbaceous chimichurri is delicious also. For non-meat eaters full-bodied flavours can be found in charred aubergine, mushrooms, lentils, roasted tomatoes and grilled peppers.

Be careful with sweetness in sauces – barbecue sauces can be matched with very ripe Californian Zinfandels but too much sweetness will accentuate a wines tannin and strip out the fruit.

Finally, older complex wines require simpler dishes – plain roast meats with their roasting juices and uncomplicated garnishes. Respect the delicacy of the wine and hold back with powerful flavours. Classic beef Wellington or simple roast game birds will pair nicely with the tertiary, mature flavours found in older wine.

Ronan's Wine Picks

CHÂTEAU PONTET-CANET, PAUILLAC, BORDEAUX, FRANCE

DOMAINE DES COMTES LAFON VILLAGE-LEVEL MEURSAULT, BURGUNDY, FRANCE

DOMAINE DUJAC, GEVREY CHAMBERTIN AUX COMBOTTES, BURGUNDY, FRANCE

SPOTTESWOODE ESTATE-GROWN CABERNET SAUVIGNON, NAPA, USA

YANGARRA OLD-VINE GRENACHE, MCLAREN VALE, SA, AUSTRALIA

Chris Howell
Cain Vineyard & Winery

3800 LANGTRY ROAD, SAINT HELENA, CA 94574, USA
WWW.CAINFIVE.COM

Wiry, soft-spoken, whip-smart, Chris Howell always seems to be about five steps ahead of me in any conversation. He has the unwavering commitment to a topic that can be unnerving if you're not fully on your game, and he is just about the furthest thing you can imagine from the Napa showmen that sometimes seem to dominate the valley.

Cain is easily one of my favourite properties in Napa, hidden up in the heights of Spring Mountain. The first time I visited it was so foggy I could barely see beyond the end of my car bonnet, so it took until round two to realize just what stunning land Chris is caretaking (as winemaker and general manager).

This is a guy who studied philosophy and critical thinking in Chicago before heading off to Montpellier in France for his first winemaking degree, then to Bordeaux to work at Château Mouton Rothschild.

Let me give you an example of how Chris never takes the easy route in wine or in conversation.

'What we have learned over the years,' he says, reflecting many of the winemakers in this book, 'is that less in more. It's easy to preach non-intervention but more difficult to find and follow the path.' Then he stops, and adds, 'of course there is intervention – we prune the vines, and we harvest the fruit.'

He just can't let himself get away with being anything other than absolutely clear.

> 'We have slowly learned the importance of a living soil. So many people have tried to explain it to me but at Cain we are slow but deliberate learners.'
> *Chris Howell*

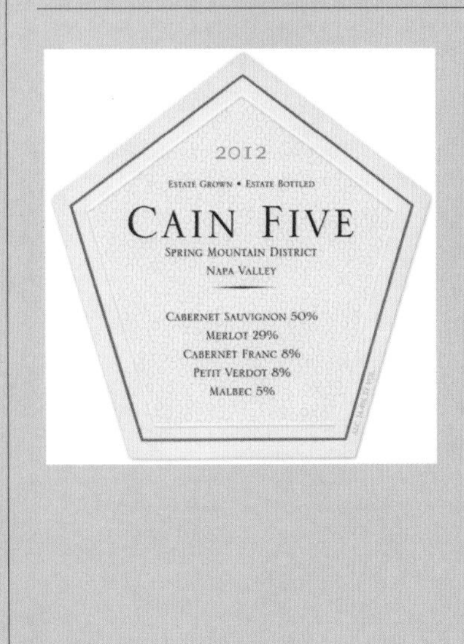

CAIN FIVE SPRING MOUNTAIN DISTRICT NAPA 2012, USA

Organic for 15 years, winemaker Chris Howell has practised biodynamics across a quarter of the vineyards since 2009, and it's clear to see that the focus of this wine is on precision, terroir-expression and balance. This is a blend of 74% Cabernet Sauvigon, 9% Malbec, 7% Cabernet Franc, 6% Merlot, 5% Petit Verdot grown on the magnificently wild slopes of Spring Mountain. Blackberry bud, savoury but not herbaceous flavours, there is a smokiness here, day-old ash with truffles and haunting florality on the finish.

'This wine should be served with a great-quality fillet steak, simply grilled and served with thyme-scented fries and roasted vine tomatoes.'
Ronan Sayburn MS, 67 Pall Mall, London

Opposite: Cain Vineyard & Winery on Spring Moutain on one of the wilder corners of California's Napa.

BENZIGER FAMILY WINERY, TRIBUTE SONOMA 2012, USA

WWW.TRIBUTEWINE.COM

It is impossible not to include the Benziger wines in this book – Mike Benziger was one of the pioneers of biodynamic farming in North America, and every wine in the group is certified sustainable, organic or biodynamic. Tribute is a blend of 67% Cabernet Sauvignon, 19% Cabernet France, 6% Malbec, 5% Petit Verdot, 3% Merlot and packs quite a punch. The spicy-plum and chocolate-infused grilled herbs of these Bordeaux grapes grown in the Californian sunshine, is wonderfully echoed with tobacco leaf and cold ash on the finish. I loved this with slow-cooked black bean enchiladas.

CLOS SARON OUT OF THE BLUE CINSAULT SIERRA FOOTHILLS 2014, USA

WWW.CLOSSARON.COM

Owned by Israeli-born, French-trained Gideon Beinstock and his wife Saron Rice, this estate is best known for its Pinot Noir, but here Gideon is working with 130-year-old Cinsault vines, blended with just a touch of Syrah. Very pale in colour, with a floral almost Pinot-like nose and lovely tart-gooseberry fruit alongside the redcurrants, wild cherries and soft spice. Vinified in open-top, old barrels, with wild yeasts from extremely low-yield vines. Not certified, but only organic treatments are used. Gideon constantly experiments and pushes boundaries. Try with Moroccan-spiced lamb.

COWHORN OREGON SYRAH 21 2013, USA

WWW.COWHORNWINE.COM

Located in Appelgate Valley AVA in southern Oregon, these guys focus on Rhône varieties that suit their rocky soils. The grapes are all estate grown alongside asparagus, cherries, lavender and hazelnut trees, and, apparently, marijuana (all perfectly legal in Oregon these days). This 100% Syrah is partly whole-bunch fermented and aged for nine months in 40% new oak. The number on the label indicates the amount of frost hours the vines were subject to from bud-break to harvest, so here 21 tells you that it was a pretty warm year. I wasn't sure about this numbering system at first, but actually I think it's a pretty good way of knowing what to expect – here the warm year means ripe damson and fig flavours, exotic spices and a rich flourish of smoke on the finish. Bring out the sweet side with a ginger, soy and garlic marinade of a grilled piece of pork or vegetables.

Opposite: Cowhorn, Oregon is certified biodynamic.
Overleaf: Benzinger Family Winery has been certified biodynamic by Demeter since 2000.

Gilles de Domingo
Cooper Mountain Vineyards

20121 SW LEONARDO LANE, BEAVERTON, OR 97007, USA
WWW.COOPERMOUNTAINWINE.COM

Gilles de Domingo's journey from France to Oregon started through a missed flight connection. He was working for the merchant company Direct Wines at the time and travelling all over the world in his role as a buyer. This particular time he was flying through Los Angeles' LAX from South Africa to New Zealand, and he had made it all the way to the departure gate, then fell asleep. Luckily he had a friend in LA who gave him a bed for the night, and when they were out for dinner he met the man – Brad Biehl of the organic King Estate in Oregon – who would soon become his new boss.

'Within a few days I was on a plane to Oregon. I didn't fall asleep this time.'

He stayed with King Estate for several years, then returned to France where his family has vineyards in Bordeaux, before deciding to move back to Oregon.

'My grandfather worked in the Spanish resistance during World War Two, helping people escape over the Pyrénées. He lost everything but eventually moved to Bordeaux where he and then my father built up a successful butcher and property business.'

'It was always in my blood to create my own story, and that has proved to be in wine.'
Gilles de Domingo

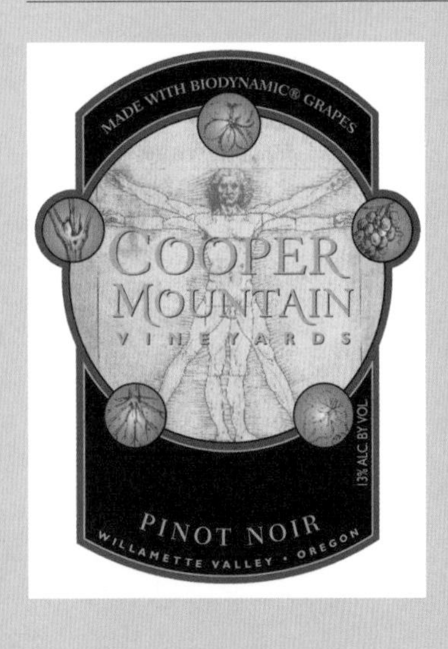

COOPER MOUNTAIN VINEYARDS JOHNSON SCHOOL PINOT NOIR WILLAMETTE VALLEY 2014, USA

This single-vineyard Pinot Noir is bursting with energy. It has a touch of sweetness on the first attack, then embraces you with ripe cherry and redcurrant. Totally moreish, it is one of three single-plot Pinot designations from Cooper Mountain and is on the Chehalem AVA, a slightly warmer vineyard site on volcanic soils, giving lovely ripeness to the fruit. Winemaker Gilles de Domingo keeps things simple – wild yeast fermentation and bottling unfiltered.

'Try a full-bodied Pinot Noir like this one, with roasted duck breast with a sauce of cinnamon-spiced cherries or with braised lentils and bacon for an earthy alternative.'
Ronan Sayburn MS, 67 Pall Mall, London

EISELE VINEYARD CABERNET SAUVIGNON NAPA 2013, USA * *

WWW.EISELEVINEYARD.COM

Easily one of the most perfectly proportioned wine estates that I have visited, you can feel the hum of a land working in balance with itself the second you step into the circle of redwood barns, beehives, apricot trees, vegetable gardens and vines. Part of the Artemis Group (Château Latour, Domaine d'Eugenie, Château Grillet); the biodynamics carried out here is reflected in their farming in Bordeaux, Burgundy and the Rhône. Effortlessly vibrant, the palate pulls you along through an array of cherry leaf, bilberry, violet, liquorice root, cedar and slate flavours with a twist of salinity. Every single element is obsessed over in the vineyards and cellar at Eisele, and it shows in the confident poise of the wine. *Confit de canard* with a green salad would be a perfect match.

HALL WINERY KATHRYN HALL CABERNET SAUVIGNON RUTHERFORD NAPA, USA

WWW.HALLWINES.COM

From the Sacrashe vineyard in Rutherford that receives huge amounts of sunshine (Craig and Kathryn Hall say it has a 'higher radiant point' than many other parts of Napa) and is set on volcanic soils. All of which makes it clear why this is so rich, structured and packed full of intense black fruit, graphite and tight, smoky caramel. The blend is 95% Cabernet Sauvignon, 4% Merlot and 1% Petit Verdot. Pair this with equally high-impact flavours, like sweet-mustard roast chicken.

INGLENOOK ESTATE RUBICON RUTHERFORD NAPA 2012, USA

WWW.INGLENOOK.COM

Francis and Eleanor Ford Coppola don't make a big deal about being organic, but the entire vineyards at Inglenook have been farmed this way in the late 70s and certified since 1991. The impetus for it came from Eleanor's organic vegetable garden, and she has been a crucial part of the philosophy being adopted across the vineyards. They grow a variety of different grapes but for the flagship Rubicon it is Cabernet Sauvignon as king. The 2012 vintage was the second under French winemaker Philippe Bascaules, who brought it a slightly softer, more sculpted style. This is still as complex, smoky and rich as you expect from an iconic Napa name, but has dark pepper, violet and grilled mint notes that tighten it through the mid-palate and keep things highly focused. Simple but striking food is called for – how about blackened tomatoes and shallots over sourdough toast?

Moe Momtazi
Maysara Wines

15765 SOUTHWEST MUDDY VALLEY ROAD,
MCMINNVILLE, OR 97128, USA
WWW.MAYSARA.COM

An Iranian immigrant, Moe Momtazi first came to the US as a civil engineering student, based in Arlington, Texas back in 1971. After graduating in 1978 he headed back to Iran, just before the Revolution began. This eventually led to Moe and his – by that point – heavily pregnant new wife fleeing the country first to Pakistan, then to Madrid before eventually making his way to Mexico and back into the US.

'I began by applying for political asylum but in the end got a job through my engineering degree and stayed in Texas for a few years before eventually moving to Oregon in 1990.' By this point he was running his own engineering business but his passion was always agriculture and in 1997 he bought the vines that would become Maysara. Today, he tells me this is the largest certified biodynamic vineyard in Oregon, and one of the largest in the US. Moe's take on organic farming is rooted in his childhood. 'My grandfather farmed the land and I have memories of walking up through orchards with him on the Caspian Sea and him talking to me about how modern farming methods would destroy our traditional agricultural system. It was a time when many people believed synthetic products offered the solution to everything in farming, but he was always concerned about manipulation of our foods.'

'When I began to farm, I thought a lot about this, and about our Persian traditions. In ancient Persia farmers were extremely sophisticated in their methods, and harnessed the energies of the planets and moon to supply the right amounts of food. I learned some of this when young, and then studied it more in depth when I started farming.'

'Of course the more I understand the more I realize that I still don't know, but I learn every day through the struggle.'
Moe Momtazi

MAYSARA WINES MOMTAZI VINEYARD OREGON 2012, USA

This is a vibrant and rich style of Pinot Noir that successfully showcases a riper, fleshier side of the grape. Moe and Flora Momtazi created the vineyard out of an abandoned wheat farm just to the south of Oregon. Located in the McMinnville AVA, it has some of the oldest soils in the state, planted with the Pommard clone of Pinot Noir, almost entirely grown from seedlings. This is a vibrant, satisfying wine that is intense and opens in the glass to reveal floral notes of violet and roses. Pair it with a hearty but simple dish like sticky barbecued chicken wings.

SPRING MOUNTAIN ELIVETTE 2013 NAPA, USA

WWW.SPRINGMOUNTAINVINEYARD.COM

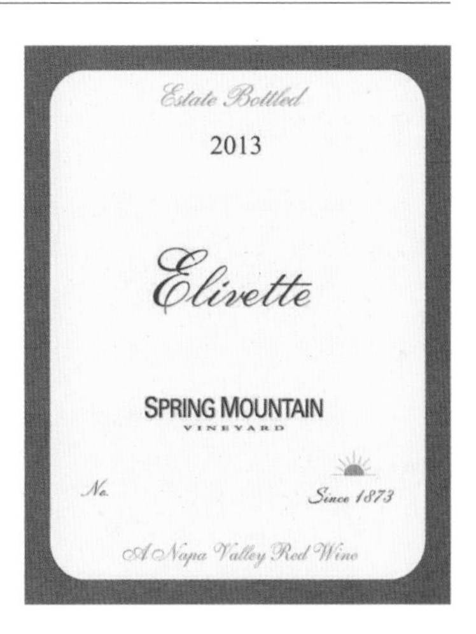

The entire Spring Mountain estate is a showcase for the benefits of biodiversity, with only 20% of its 342 ha given over to vines, and the rest left to wild land, filled with native trees such as oak, madrone, pine and redwoods... and a ton of wildlife, which all play a part in keeping this ecosystem in balance. Elivette is the wine most clearly aimed at long ageing, filled with fine tannins and elegant, restrained fruit. Comprised of 80% Cabernet Sauvignon, with a rounding out of Cabernet Franc, Merlot and Petit Verdot, it resonates through the palate with liquorice, mint, bilberry, cassis and chocolate. This is still young and fresh, with an attractive astringency to the tannins on the close of play. They are really cradling this fruit, digging in for the long haul. Tea-smoked duck would be a stunning match to this.

SPOTTSWOODE FAMILY ESTATE-GROWN CABERNET NAPA 2013, USA * * *

WWW.SPOTTSWOODE.COM

Brilliant wine from one of my favourite producers: wild strawberry and cherry dominate, with structure and hold from the tannins and softer tertiary flavours of tobacco and liquorice starting to come through, while still offering rich fruit. The epitome of power and restraint rolled up into one. Organic since 1985, the commitment is clear as Aron Weinkfauf takes charge of both viticulture and winemaking – so often seen as two separate areas in Napa, but not at Spottswoode.

'Serve with good-quality venison and a red wine sauce with blackberries and a touch of truffle oil.'
Ronan Sayburn MS, 67 Pall Mall, London

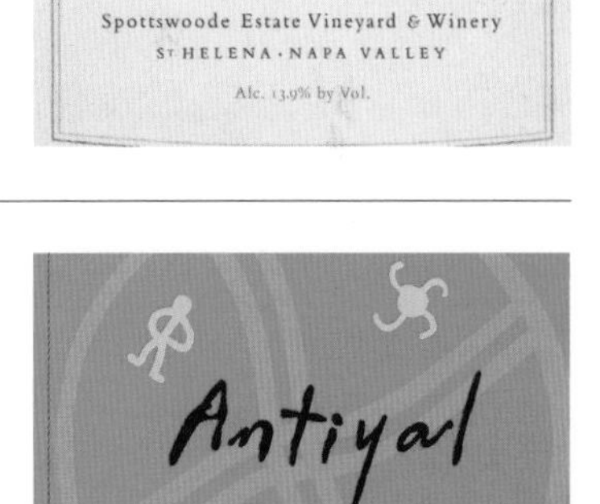

ANTIYAL MAIPO 2012, CHILE

WWW.ANTIYAL.COM

Sometimes referred to as Chile's first 'garage wine', Antiyal is owned by Alvaro Espinoza, and the first vintage was in 1998. This wine is a blend of 49% Carmenère, 36% Cabernet Sauvignon, 15% Syrah, farmed with minimum irrigation and naturally low yields from two different vineyards in Maipo. All composts are made from the family's own farm. Expect gorgeously rich and vibrant flavours.

'Delicious nose with abundance of black fruit, sweet spices, roasted coffee followed by a well-structured palate. It is a great wine that can age well and that would go superbly with a roasted rack of lamb with a spicy pumpkin purée.'
Gerard Basset MW, Best Sommelier in the World 2010

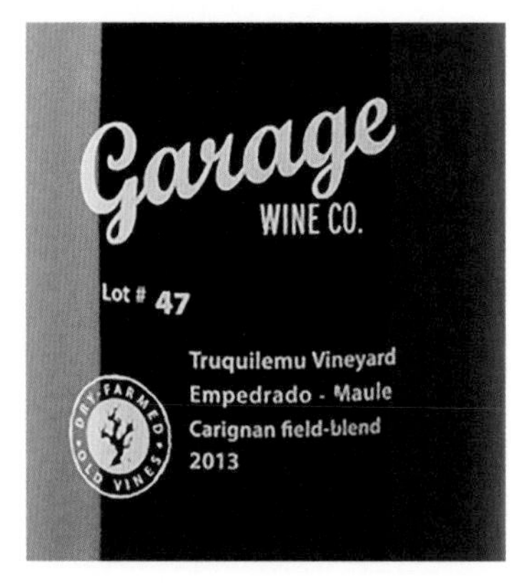

GARAGE WINE CO TRUQUILEMU VINEYARD CARIGNAN/MATARO MAULE 2013, CHILE

WWW.GARAGEWINECO.CL

These wines are not certified, because all the Garage Wine Co growers are small-scale, traditional farmers. But owner Derek Mossman's response to the question is perfect: 'are our vineyards organic? Absolutely, and not just from when we sought a marketing opportunity, but for many many generations, since our growers' forefathers tended these wines with horses, sweat and small amounts of sulphur.' The 70-year-old Carignan vines are dry-farmed, fermented in small open-top tanks, then aged in barrel (no new oak) for over two years. Smoky, earthy flavours abound, with dried herb notes curling up through the red fruits. Tiny quantities, but all the wines in the range are made in the same way. You want deep flavours in your food matching for this one – a spiced-chicken casserole with wild rice would be delicious.

SEÑA ACONCAGUA 2014, CHILE *

WWW.SENA.CL

Owned by Edward Chadwick of Viña Errázuriz, Seña has been biodynamic since 2004, and gradually the amount of plots coming from the Errázuriz Colchagua vineyards has been dropped (but the winemaker reserves discretion on this, and has the final call each year). This wine is a blend of 60% Cabernet Sauvignon, 16% Carmenère, 11% Malbec, 8% Merlot and 5% Petit Verdot: packed with fresh but intense cardamon and black-pepper spice and beautifully integrated, intense but silky tannins. Smoke, blackcurrant and caramel notes appear after a few minutes in the glass. Needs another few years in bottle to really limber up. One of the Seña winemaking team suggests a pairing of Patagonian lamb, barbecued with a little fresh rosemary sauce.

BODEGA COLOMÉ MALBEC AUTENTICO, ARGENTINA 2014

WWW.BODEGACOLOME.COM

Located in the upper Calchaqui Valley, this is one of the highest vine-growing areas in the world. Part of the Hess Family Wine Estates, Colomé used to be certified biodynamic but no longer is, although all vineyard practices remain the same for the Colomé wines (but not for their La Brava vineyard in Cafayate). The vines here have an average age of 90 years, and are aged in cement tanks to preserve the fruit emphasis. You still get the intense, rich spiciness, but the floral side of the grape is emphasized, as is the acidity, making this a brilliant food wine to match with a lean cut of beef, topped with the traditional Argentine chimichurri sauce.

Opposite: Bodega Colomé in Salta, Artentina, has vines that are 160 years old.

BODEGA NOEMIA MALBEC RÍO NEGRO 2014, ARGENTINA

WWW.BODEGANOEMIA.COM

Seriously great Malbec from this Danish-Italian winery in the Río Negro Valley of Patagonia – a partnership between Countess Noemi Marone Cinzano and Hans Vinding-Diers. Concentrated liquorice with juicy Black Mission figs, tobacco notes, crème caramel and spiced chocolate; the alcohol is high and oak unmistakeable, but there is focus, lift and tension and this is seriously easy to sink into. Everything is done by hand here, including the bottling.

'This elegant wine comes from rugged and remote Patagonia – it might be tricky to do at home but chefs down there roast butterflied goats whole over open charcoal pits and serve with herbaceous and garlicky chimichurri sauce.'

Ronan Sayburn MS, 67 Pall Mall, London

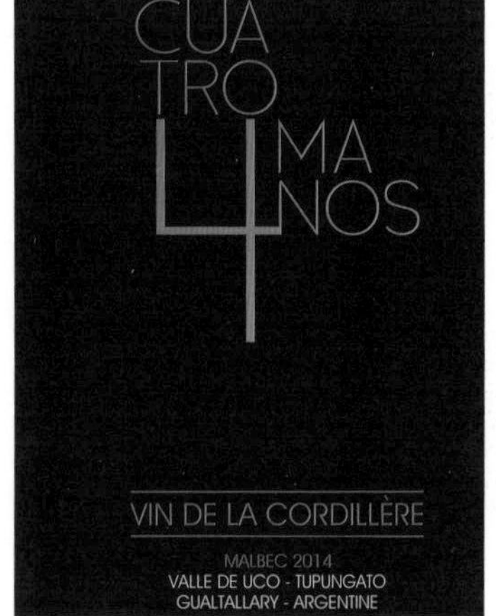

CUATRO MANOS VINO DE LA CORDILLÈRE MALBEC TUPANGATO 2014, ARGENTINA

WWW.FACEBOOK.COM/WALLARD.VINCENT

Vincent Wallard makes wine in Argentina and also in Chouzé-sur-Loire in Saumur. He's laid-back and intuitive – questions about sulphur levels, for example, are met with, 'I bottle some with, some without, depending on how I feel the wine is responding, and so I can follow its progression.' This is made from grapes grown at 1400m (4593ft) altitude in Tupangato, vinified in concrete tanks with 30% whole bunches and the rest destemmed. The wine has soft notes, hand-crushed raspberry purée and aromatics of wild *garrigue*: a great match to a smoky black bean burger with a red cabbage salad.

DOMAINE BOUSQUET GAIA RED BLEND TUPANGATO 2013, ARGENTINA

WWW.DOMAINEBOUSQUET.COM

The Bousquet family is the number one exporter of organic wine in Argentina. Although they been here since the late 90s (moving from France's Langeudoc), 2013 is the first vintage of this wine. The vineyard sits at 1200m (3937ft) altitude, and the blend is 50% Malbec, 45% Syrah and 5% Cabernet Sauvignon. On the nose, it's Syrah that wins, with a floral, fragrant hit followed up by an intense core of spice. Tannins have the soft and supple character that is one of the real joys of Argentina. Succulent and flatteringly fruit-forward blackberry and olive flavours. The 80% new-oak ageing requires bold flavours: burritos or burgers, chilli con carne or even a roasted beetroot salad.

ÁLVARO PALACIOS VI DE VILA GRATALLOPS PRIORAT DOCA 2014, SPAIN

WWW.ALVAROPALACIOS.COM

Owner and winemaker Álvaro Palacios is one of the legends of Spain. Here he has produced a blend of 80% Garnacha (Grenache), 20% Cariñena (Carignan), with the alcohol so perfectly balanced by the low pH3.2 that you can practically walk along the tightrope of this wine. Unfilted and unfined, it is still in the earliest stages of its life, and ideally should be put away for another five to ten years. Inky, dark-purple in colour, this pretty much draws a picture of what it means to say; a wine with integrity in its fruit structure. Fleshy blackcurrants and touches of chocolate, ginger and mint, concentrated but soaring freshness on the finish. Gratallops is a village, the grapes are located on steep terraced slopes, worked by mules. Match with paprika, garlic, olive oil, sweet peppers, liberally served with vegetables or meat.

ARTADI VALDEGINES RIOJA ALAVESA 2013, SPAIN

WWW.ARTADI.COM

Chocolate and black cherry on the attack gives this a rich and polished feel, with graphite and floral aromatics. From the Tempranillo grape, there is a sense of high drama here, with well-worked tannins that will hold on for years. Winemaker Juan Carlos López de Lacalle works on site-specific wines, unlike much of Rioja, and has been farming organically for 40 years, saying he has never even bought a fertilizer. Instead he ploughs naturally growing herbs and flowers into his soils, from poppies to alfalfa, mustard and thistle. These vines are grown in a small valley called Rio San Gines, and produce among the most fruit-forward wines of Artadi's single vineyards. Chocolate, black cherry and floral aromatics would work well with a wild boar stew.

DOMINIO DEL ÁGUILA PÍCARO DEL ÁGUILA TINTO RIBERO DEL DUERO 2012, SPAIN

WWW.DOMINIODELAGUILA.COM

A new estate that is making all the right impact. Founded by Jorge Monzon in 2010 (it also has a micro-brewery) when he renovated a 17th-century winery with underground cellars dating right back to the 15th century. Natural-yeast fermentation, foot pressing and indigenous grapes – primarily Tempranillo, but with touches of Bobal, Blanca del Pais, Garnacha, Tempranillo Gris and Albillo; aged for 19 months in oak. There is so much depth of blackberry fruit – so pure that you could almost reach into the glass and pick out the berries – with depth and complexity from liquorice root and bitter chocolate. Unfiltered and unfined, this is made for spicy chorizo in any form.

Peter Sisseck
Dominio de Pingus

HOSPITAL, S/N 47350 QUINTANILLA DE ONÉSIMO
(VALLADOLID), SPAIN
WWW.PINGUS.ES

One of the most amazing things about wine is that the biggest names, from the most iconic estates, are often the most down to earth and easy to talk to of all producers. Peter Sisseck is absolute proof of this.

Pingus stands among the most iconic wines of the world, but its creator (its named after his childhood nickname) is someone who never ceases to question what he's doing. It's a trait that is at the heart of his success.

'Back in 1983, I worked in Bordeaux for one year during a challenging harvest and saw that systemic treatments did not always mean success. It gave me a firm belief that you have to be cautious and not always listen to people trying to sell you products. Then, back in Denmark I studied agricultural farming and I met a girl – she was pretty nice now I remember – who showed me a really interesting trial about how long biodynamic fruit lasted compared to those farmed conventionally. I thought was pretty wild! Then I tried to read a bit of Steiner and like most people didn't understand it.'

'But the more I read about the winemakers following biodynamics, the more I loved the sheer insanity of keeping going through the difficulties because of the results in the wine, and I just knew it was what I wanted to do.'

> 'I love the basic simplicity in the idea that as a farmer, the soils are your number one capital asset – and if you don't keep them healthy; you are spending your capital.'
> *Peter Sisseck*

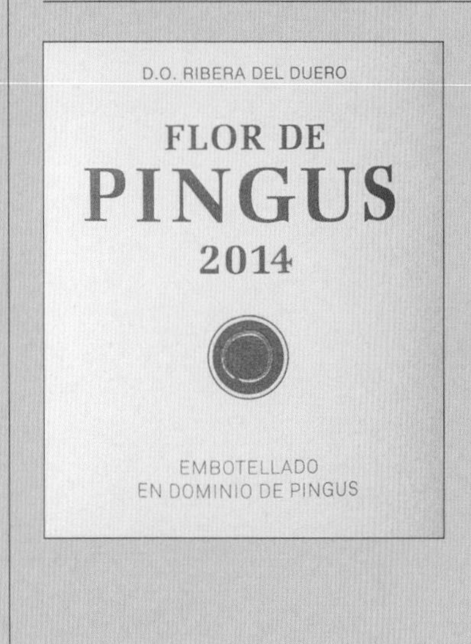

FLOR DE PINGUS RIBERO DEL DUERO 2014, SPAIN

I'm not sure that any wine has ever stopped me in my tracks the way Pingus did; it was the fresh mint notes that I remember, giving a turbo-charged thrill to the intense black fruits. This is the second wine, Flor de Pingus, still 100% Tempranillo, but from vines that are over 35 years old, around half the age of those in Pingus, and aged for 19 months in one-year-old French oak barrels. It approaches the complexity, intensity and ability to surprise that its big brother does. Danish owner Peter Sisseck has been farming biodynamically since 2000, and the vitality and sense of forward motion that he manages to capture in his wines should be experienced. Expect pure, intense black cherry, with rich liquorice and aniseed overtones. A perfect match for *habitas con calamares* – a flavoursome stew of fava beans and squid.

Opposite: In the running for the most iconic vineyard land in Spain: Pingus.

Miguel Torres Senior
Bodegas Torres

6. 08720 VILAFRANCA DEL PENEDÈS, BARCELONA, SPAIN
WWW.TORRES.ES/EN/WINES/MAS-LA-PLANA

I challenge anyone to spend time with Miguel Torres Senior and not come away feeling both humbled and enriched by the experience. This is someone who runs a business empire of over 2000 ha of vines and more than 1300 employees and yet who has shown both serious and sustained commitment to the environment.

Among his many initiatives has been the planting of indigenous varieties and the reviving of ancestral grapes in his vineyards, requiring less intervention in vineyard and cellar and therefore minimal use of additives. He has also been a world leader in purchasing land in regions that are today at the outer limit of possible planting zones for vine ripeness, but that he believes will soon be perfectly adapted to winemaking, so drawing attention to the dangers that climate change poses to the regions currently making wine.

He is a staunch advocate of organic farming. Torres has 600 ha of certified organic vineyards in Spain, 350 in Chile and 32 in California. In Chile certification was made in 2012; the rest are in conversion, if not there yet.

'Although my belief is that organic is not enough,' he says in his quietly passionate manner. 'I hope in the future organic wines take climate change into consideration – there are today far more things to consider that go beyond today's regulations. I want us to think about using ecological ink for printing labels, using bottle caps free of heavy metals, using only recyclable glass and to compost all vineyard waste.'

'Adapting to climate change is the true challenge of the 21st century.'
Miguel Torres Senior

MIGUEL TORRES MAS LA PLANA PENEDÈS 2011, SPAIN

Miguel Torres Senior's emblematic estate of Mas La Plana covers 29 ha in Penedès. This is one of the wines that I hesitated over, because as 100% Cabernet Sauvignon it hardly promotes indigenous grapes. And yet, close to its 40th harvest, it has become an icon of a man who has done more to raise awareness of climate change than almost any other in the wine industry. A supremely sophisticated bottle that is still gloriously young and will another five years before it starts to fully soften, it is full of cassis and truffle, but with an extra intensity of toast, peppery sage and laurel notes, wrapped up in gentle yet insistent tannins. A brilliant food wine, give it a well-braised short rib.

Above: Miguel Torres of Mas La Plana has lobbied for a greener wine industry for decades.

Roberto Olivan
Tentenublo Wines

C/LA FUENTE 52-54, 01308 LANCIEGO (ÁLAVA), SPAIN
WWW.TENTENUBLO.COM

Roberto is one of the winemakers that made me want to write this book. His wines are so beautiful, and his own attitude so refreshing (he routinely gets described as irreverent, an iconoclast or even the reincarnation of Maximilien Robespiere or Alexander Hamilton!), that to me he pretty much sums up the spirit of artisan winemaking.

Roberto created the winery himself back in 2011 from plots left to him by his mother and grandmother at various times, some when he was 16 and some when he turned 18 years old. His label designers gave him the job of coming up with a striking name and he went with Tentenublo, which was the word given to the bell ringing used to fight off hailstorms in some Rioja villages. He farms organically (although he says 'I have no time for official certification'), does his own massal selection, uses only natural yeasts, never acidifies or uses any other additives except extremely low quantities of sulphur, foot stomps the vines and is a loud proponent of terroir-driven, single-vineyard Rioja. His Escondite del Ardacho vines are labelled by single-vineyard names, something that has been the subject of fierce debate for years and that the local regulatory body only officially recognize in June 2017.

He is open and willing to chat to anyone about his wines; if you come to him, as he doesn't like to travel. 'You have to remain in one place,' is how he puts it.

TENTENUBLO WINES TINTO ROSSO RIOJA DOCA 2015, SPAIN

When you drink Robert Olivan's wines, you enter his world – full of swagger. This is tight and young right now, but wow is it good. Silky and ready to go after a few hours in a decanter, packed full of beautiful red fruits from the Tempranillo and Garnacha grapes grown on poor soils with no more than 50cm (20in) of dirt before they hit bedrock. Bottled with extremely low doses of sulphur, this is as complex as you would hope for from vines that are aged between 40 to 100 years old. This is a winemaker and a label with a great future. Food-wise, you just know a good plate of cured Spanish ham is going to do the business.

FAMÍLÍA NIN-ORTIZ PLANETES DE NIN PRIORAT 2013, SPAIN

CARLESOV@GMAIL.COM

Oh but this is lovely. So focused and intense right from the word go, with layers of olive tapenade and dripping with fully ripe bilberry and damson fruit, wrapped in heavily spiced cinnamon toast. Winemaker Ester Nin manages to make these tannins both burly and yet shyly sweet. This is vinified with wild yeast in large oak vats or in amphora, where they are also aged. The striking thing is how pure and balanced this is, even with its undeniable intensity. It seemed to get lighter in my mouth over the tasting, a testament to the gossamer quality of the tannins. Keep the food intense but simple also – miso-marinated black cod would be perfect.

SUERTES DEL MARQUÉS, LA SOLANA TINTO, VALLE DE LA OROTAVA, TENERIFE 2013, SPAIN

WWW.SUERTESDELMARQUES.COM

Jonathan Garcia's wines are making all kinds of waves, as well they should, showcasing just what brilliant things are happening in the Canary Islands. With a focus on organically farmed indigenous grapes and traditional vine-training methods, the Solana Tinto is focused and powerful, while still having a freshness and clarity to the dark fruit flavours. It is made with 100% ungrafted Listán Negro grapes (some of the vines are up to 200 years old), fermented in concrete, then with one year in large 500-litre barrels. Natural yeasts, low sulphur, with an unmistakable minerality from the volcanic slopes of Mount Teide. Find this wine, sit down with a plate of something hot and spicy and enjoy.

CHÂTEAU FONROQUE ST-ÉMILION GRAND CRU CLASSÉ 2012, FRANCE

WWW.CHATEAUFONROQUE.COM

Although Pontet-Canet gets a lot of the attention for introducing Bordeaux to the benefits of biodynamics, Alain Moueix over at Fonroque deserves his fair share of credit. For Alain, one of the keys is that biodynamics brings phenolic ripeness to the grapes earlier than conventional farming does, which gives him leeway for picking in difficult vintages. I love how his wines stand out in St-Émilion for their freshness. This 2012 is a perfect example – juicy frame, violet and raspberry aromatics, pencil-lead, elegance but with plenty of plummy fruit impact. This is a blend of 85% Merlot and 15% Cabernet Franc; aged 20% in cement tanks, 80% in oak barrels with half of them new. An easy match to a lamb stew with roasted root vegetables.

Bapiste Guinaudeau

Château Lafleur

33500 POMEROL, FRANCE
WWW.CHATEAU-LAFLEUR.FR

'We are suspicious of doing anything too quickly, because the margin for progression in great wines is in the nuance. It's tempting to think everything needs to be black or white in decision making, but the truth is usually somewhere in between.'
Baptiste Guinaudeau

Baptiste applies this reasoning to everything, even organics. 'We always look at what makes simple good sense. We try to honour the common sense that our ancestors applied before us – and the good luck that sometimes comes with it. For example one of the reasons that this estate is farmed so traditionally, with such old vine stock, is because when my great aunts were in charge from 1947 until 1985, they were extremely conservative, so when the trend was to plant modern Merlot clones and use chemical weedkillers, they were reluctant or unwilling to follow suit. And then when my parents took over they had no money, so also couldn't do any modernizing. Today I owe them all a huge favour.'

Baptiste runs Lafleur, the tiny Pomerol estate whose name quickens the heart of wine-lovers world over, with his wife Julie. Together they look after the 4.58 ha not by row of vines but by individual plants, with yields measured by the number of glasses and bottles that the fruit of each of the 21,000 vines will give.

CHÂTEAU LAFLEUR POMEROL 2014, FRANCE * * *

The nose of this wine is spicy, cedar and white pepper with an obvious Cabernet Franc edge (it makes up 55% of the blend, with Merlot the other 45%). The attack is intense, with black olives, tapenade and raspberry leaf, all bristling with intent. There is a juicy tension to the structure that creates a sense of drama, and although it is extremely young right now, you can feel the layers building a sense of complexity. After half an hour in the glass the violet aromatics start to unfold. Give this time, and you will be richly rewarded. Not certified organic, but no chemicals, pesticides or weedkillers are used. Baptiste is clear to underline he doesn't follow any system religiously. Truffles always work with Pomerol – shave them over pasta or a rich beef stew as you wish.

Above: Lafleur in Pomerol is artisan winemaking at its best.

Jean-Pierre Amoreau
Château le Puy

33570 SAINT-CIBARD, FRANCE
WWW.CHATEAU-LE-PUY.COM/EN

Jean-Pierre Amoreau is a man who positively relishes his outsider status in Bordeaux. There has been a member of his family making wine just outside the village of Saint-Cibard since 1610, and yet the idea of being part of the establishment invariably raises a huge laugh.

> 'The system of Bordeaux is run according to appellations and so rewards conformity. We have no issue with appellations, but just believe that our wine tastes different from our neighbours and so needs one of its own.'
> *Jean-Pierre Amoreau*

Never one to stand on ceremony, this means that Jean-Pierre together with his children Pascal and Valérie have created an appellation of their own.

The Le Puy charter (as yet unaccepted by France's official governing body, it should be noted) details that this is a biodynamic-only AC, with 1 ha of wild flowers or hedgerows conserved for every hectare of vines. Chaptalization is not allowed and all harvesting is by hand, with no green harvesting permitted; natural-yeast fermentation and bottling at the château only during a full moon with either no or minimal added sulphur. Wine ageing takes place with no new oak and the barrels used are between five and fifty years old, in complete opposition to most Bordeaux châteaux, which use new oak, or at most two or three year old.

It's ambitious, reassuringly thorough, slightly bonkers and totally admirable – which pretty much sums up Jean-Pierre in my experience.

CHÂTEAU LE PUY CUVÉE BARTHÉLEMY FRANCS-CÔTES DE BORDEAUX 2010, FRANCE

Jean-Pierre Amoreau is the original anti-establishment Bordeaux winemaker. He runs one of the few estates in Bordeaux to work closely with Claude and Lydia Bourguignon, the terroir and soil experts. The specific plot for this wine is called Les Rocs, which has a distinctive limestone soil, giving a tickle of minerality to the finish. The wine is a blend of 85% Merlot and 15% Cabernet Sauvignon and is all vivacity, rich fruit and authenticity. It is aged for 24 months in old oak barrels, and is lovely match for a mushroom and game pie.

Thomas Duroux
Château Palmer

33460 MARGAUX, FRANCE
WWW.CHATEAU-PALMER.COM

No one can doubt Thomas Duroux's fine wine credentials, he made wine in Hungary, South Africa, California, Tuscany, Bordeaux, and was winemaker at the legendary Tenuta dell'Ornellaia before being heading to Château Palmer in 2004 at the age of 34. So, there is something very convincing when he says,

'I believe there is now no other choice if you want to produce quality wines, you have to avoid chemicals. The progression towards biodynamics is hard, even frightening at times, but I no longer feel like an engineer of great wine. Instead I feel part of something bigger.'

Thomas has a French father and an Italian mother, and speaks perfect English. He argues fluently and openly for the reasons he decided to take Château Palmer – a bastion of classified Bordeaux – towards biodynamic farming, starting slowly in 2008 then converting over the entire property from 2011.

'We saw the changes in the vineyard, the structure of the soil deepen and thicken. But at the same time the fruit in the wine was becoming clearer. It had a definition that I hadn't seen before. I had to convince the shareholders, but luckily the idea of lower yields was never an issue because at Château Palmer we have always been about high level viticulture. Our goal was never high yields.'

'The really important thing for us is the effect on the ecosystem, the idea of preserving this piece of land. Now we have eight cows and eighty sheep, and have planted fig, apple and cherry trees widely throughout the vineyard to complement our forty hectares of forest. We're just at the beginning.'
Thomas Duroux

CHÂTEAU PALMER MARGAUX GRAND CRU CLASSÉ 2014, FRANCE * *

Palmer is always a stunning wine, and the definition and focus here is amazing; the 2014 vintage had a cool summer but excellent end of season sun. This wine has hugely dense yet fluid plush blackberry, plum and damson fruit, from a blend of 45% Merlot, 49% Cabernet Sauvignon 49% and 6% Petit Verdot. This wine moves so perfectly, you feel it ripple through your mouth. The floral edging becomes clearer after a few minutes in the glass, as does the salinity on the finish. This will not even begin to reach its peak for another decade. A leg of lamb served with potato gratin would be a classic match.

Jean-Michel Comme
Pontet-Canet

33250 PAUILLAC, FRANCE
WWW.PONTET-CANET.COM/EN

Jean-Michel was probably my first exposure, close up, to biodynamic farming. Back then it was really something new in Bordeaux, and Pontet-Canet was unwilling to talk about it too publically. But there was no hiding the revolution that was happening in the bottle, and slowly but surely they have become a world-leading proponent of biodynamic viticulture.

This is in no small part down to the estate director and winemaker Jean-Michel Comme. I suggest clearing a few hours in your diary if you are lucky enough to get to meet Jean-Michel. If you get him on to the subject of biodynamics, he can talk the hind legs off a donkey (and yes, there are a couple of those lurking around what is increasingly becoming Pontet-Canet Vineyard and Farm).

'A vineyard, its ecosystem and its response to different vintages is always a balance between four points,' he said in his deliberate-and-measured way the last time we met, drawing out a cross on whatever piece of paper he had to hand.

'Air and Earth are at opposing points of one axis, then Water and Fire the opposing points of another axis.'
Jean-Michel Comme

'Biodynamics aims to provide balance, so if there is too much Air in a vineyard site, which represents elegance, the treatments will aim to provide Earth to bring depth and intensity to the wine. Similarly if a wine has too much Fire, or alcohol, then the treatments aim to bring Water, or drinkability. Great wine has a balance between all four elements. It is not a complicated idea at its heart, and yet we tie ourselves up in knots over it at times.'

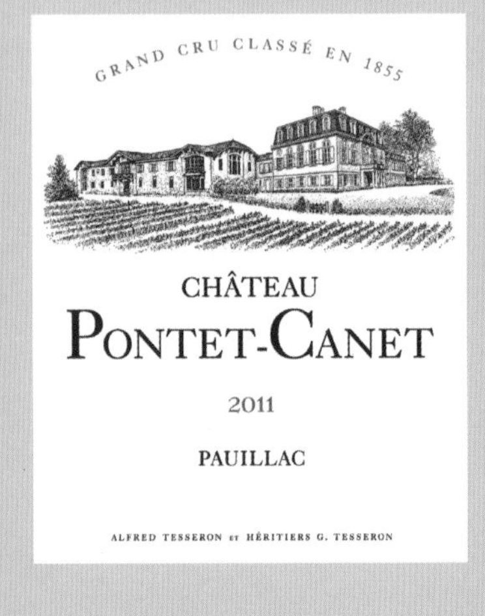

CHÂTEAU PONTET-CANET PAUILLAC GRAND CRU CLASSÉ 2011, FRANCE *

This is a wine to age, of course, but the 2011 is a lighter-style than some vintages, and the subtleties are already starting to be revealed. High aromatics, vibrant cassis and bilberry-fruit is immediately apparent; the flavours build slowly but surely over several minutes. Even with the firm tannic structure, there is clear minerality and floral tones, all underpinned by layers of black truffles and liquorice.

'A classic beef Wellington with a few extra slices of black truffle under the pastry and sauce Bordelais would be a wonderfully elegant accompaniment with this wine.'
Ronan Sayburn MS, 67 Pall Mall, London

CHÊNE BLEU HÉLOÏSE 2010 IGP VAUCLUSE, FRANCE *

WWW.CHENEBLEU.COM

Chêne Bleu is located high up, at around 550m (1804ft) altitude, in the Dentelles de Montmirail mountains in the Southern Rhône. It is named after an ancient oak tree on the property that was damaged by storms and subsequently painted blue with *bouillie Bordelaise*, a treatment typically used to protect vines, by artist and sculptor Marco Nucera. Nicole and Xavier Rolet bought the estate, then called Domaine de la Verrière, in 1996 with Xavier's sister Bénédicte and her husband Jean-Louis Gallucci and set about entirely restoring the vineyards, house and surrounding pine and oak forest. UNESCO has designated the region a Biosphere Reserve. The Héloïse bottling (yes, there is also an Abélard) is a classic Rhône blend of Syrah and Grenache but with a touch of Viognier to add fragrance and lift. The definition of a rich, sculpted, fragrant, yet fresh red. The estate provides Provençal bento boxes for visitors, packed with olives, hams, fresh vegetables and local breads – try re-creating this at home.

Above/Overleaf: Chêne Bleu sits high in the mountains of the Rhône.

Sébastien Vincenti
Domaine de Fondrèche

84380 MAZAN, FRANCE
WWW.FONDRECHE.COM

I first got in contact with Sébastien Vincenti when he was creating something of a stir by withdrawing from official Ecocert organic certification in order to 'remain coherent with my organic philosophy', specifically applying certain synthesized products that are still chemical-free but not allowed within the organic certification. He's an example of how no system is perfect – and no one who tastes his wines can doubt his commitment to excellence, or to working in a way that minimizes his impact on the environment.

Sébastien joined his father Nanou Bartélemy in 1993, fresh from his oenology degree, and ever since has set about teasing the best out of the rocky plateau of Mazan in the Ventoux . He set about first farming with organics, then biodynamics, then questioned the whole thing. In the interests of balance, I think it's only fair that he gets his say here.

'Are the natural products used in organics better for the environment? This is what I have increasingly begun to ask myself. It can be better for the environment perhaps to use certain synthesized products that build up the plant's natural defences but that are not allowed under organic rules.'

> 'I became organic by conviction, but today I feel there is a big debate needed that not enough people are willing to engage with.'
> *Sébastien Vincenti*

DOMAINE DE FONDRÈCHE DIVERGENTE VENTOUX 2015, FRANCE

Winemaker and owner Sébastien Vincenti apprenticed under Rhône-legend André Brunel in Châteauneuf-du-Pape before joining his parents at Fondrèche. Juicy, big, powerful, damson colour, rich and round, throbbing with life, this is a full-on classic Syrah-led Rhône (this cuvée bottles the estate's oldest Syrah vines). Silky and fruit forward, clear oak markers, but not overdone; a true supper wine, totally versatile with hearty stews or homemade pizza. There is a great focus on the finish here, with slight point of retraction that gets your mouth watering and ready for more.

CLOS DE JAUGUEYRON HAUT-MÉDOC 2014, FRANCE

45 RUE DE GUITON, 33460 ARSAC

2014
Clos du Jaugueyron
HAUT-MÉDOC
Appellation Haut-Médoc Contrôlée
750 ML
13 % Vol.
Michel THÉRON Viticulteur
à Arsac 33460 Margaux - France
Mis en bouteille à la propriété
PRODUCE OF FRANCE
CONTAINS SULPHITES

Winemaker Michel Theron has owned this tiny estate with his wife Stéphanie since 1993, and produces wine each year from vines in both the Margaux and Haut Médoc appellations. Michel is originally from the Languedoc, which is perhaps why he has never had any qualms about approaching things in Bordeaux entirely in his own way. This is so young and tight right now that you really need to leave it for another good few years. Heartfelt Bordeaux, brilliant, classic claret, with that mouthwatering touch of fruit austerity and a sense that the tannins are just starting to merge into it, stretching and softening both elements of the wine. Wild mushrooms on toast serves me well with this wine.

DOMAINE DE VILLENEUVE LES VIEILLES VIGNES CHÂTEAUNEUF-DU-PAPE 2012, FRANCE

WWW.DOMAINE-DE-VILLENEUVE.FR

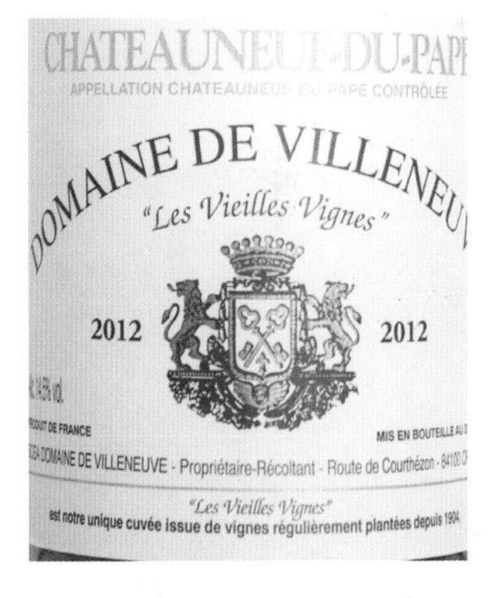

One of the founder members of Nicolas Joly's 'Renaissance des Appellations', Domaine de Villeneuve is owned by Stanislas Wallut. Low-key but a real insider's wine, this is classic Châteauneuf-du-Pape. It's hard to see which part of the vineyard doesn't qualify for 'Vieilles Vignes' really, as 90% of the estate has vines aged between 30 and 95 years old. This is a blend of 70% Grenache, with touches of Mourvèdre, Syrah, Cinsault and Clairette and is a wine to age and to savour, which is why I have picked a slightly older vintage. At this age you start to see the dark *garrigue* spices, supple tannins and multi-textured autumnal fruits that make this so complex and rewarding. Linger over it with a slow-cooked stew. Also worth looking out for their brilliant Côtes de Rhône called Cuvée La Griffe.

DOMAINE DU CLOS DES TOURELLES GIGONDAS 2014, FRANCE *

WWW.FAMILLEPERRIN.COM

A lesser-known estate from the Perrin family of Château Beaucastel in Châteauneuf-du-Pape. Organic since the 50s, and biodynamic since the 80s at Beaucastel, they bought Clos des Tourelles in 2008. A blend of 80% Grenache, and the rest Syrah, this is young and tight right now, lay it down for a few years if you can possibly resist. It's rich, deep, textured, with a buttress of a body that you smack against but then it melts away; lashings of black chocolate, olive and liquorice and dark brambly fruit. The vines (some ungrafted from the 1890s) are located in a windy and relatively cool position, sheltered by the Clos walls. The nearby family restaurant L'Oustalet is a must try.

Gérard Bertrand
Clos d'Ora

MINERVOIS LA LIVINIÈRE, FRANCE
WWW.GERARD-BERTRAND.COM

A few years ago I was translating Gérard Bertrand's book, *Wine, Moon and Stars*, and we arranged to meet up in a bistro in Bordeaux to talk through his philosophy of winemaking. This is not someone who walks into a restaurant unnoticed. A former professional rugby player, Gérard towers above pretty much anyone else in the room. And he combines his height with the kind of confidence that comes from being seriously successful. But then you speak to him and he has the soul of a poet. He talks about the spiritual dimension of wine, and the essential truths that we can find within nature.

After a while, you start to see just how he has achieved the seemingly crazy feat of converting over 450 ha of vines to certified biodynamic farming, across ten different estates in a variety of Languedoc appellations. This makes him the largest biodynamic wine producer in France (he has a further 200 ha of vines that are in conversion, so we can't fault him for effort).

'Just seeing how organic food and wine has been accepted by supermarkets gives me hope for humanity, but biodynamics goes further. It brings an awareness of the spiritual dimension of wine, and binds us together with nature.'
Gérard Bertrand

GÉRARD BERTRAND CLOS D'ORA MINERVOIS LA LIVINIÈRE 2013, FRANCE

Gérard Bertrand is the biggest biodynamic producer in France, and a passionate advocate of its benefits. This is the wine that he sees as the centre of it all; a former sheep farm enclosed by dry-stone walls in the Minervois hills, he bought it in 1997. Richly powerful, it ideally needs another five years+ in bottle. I did a double decant and then gave it time in the glass. This is a physical wine, in that it has sides and depth. A blend of Syrah, Mouvèdre, Carignan and Grenache, there are clear herbal steaks of rosemary and *garrigue* through the body, coupled with olive tapenade, scrub, bitter chocolate and intense black fruits. I tasted it first with a well aged Parmesan and it was an excellent match, but the next night I went with a chargrilled steak. By this point the wine was more open, and it cut right through the richness of the steak, totally mouthwatering.

'If you are quick enough to catch one of the local boars that sometimes roam the vineyards of the Languedoc, then slow braising it with a mix of root vegetables and a bottle of Minervois would make a welcoming dinner during the winter months.'
Ronan Sayburn MS, 67 Pall Mall, London

LE DOMAINE MONTIRIUS TERRE DES AÎNÉS GIGONDAS 2012, FRANCE

WWW.MONTIRIUS.COM

Biodynamic since 1996, Montirius makes wines across Vacqueyras, Gigondas and the Côtes de Rhône, with all 58 ha farmed holistically by Eric and Christine Saurel, today helped by their daughters (Montirius is very sweetly a contraction of their names Manon, Justine and Marius). The Terre des Aînés bottling is a blend of 80% Grenache and 20% Mouvèdre, aged for 18 months in cement tanks with no oak. Even so, this is inky and concentrated, barely out of starting blocks at five years old. Smoky bacon and *garrigue*, with crushed mint leaves on the finish. Pair with teriyaki beef or some other slightly smoked, intensely flavoured meat or vegetable.

M. CHAPOUTIER LE PAVILLON ERMITAGE 2012, FRANCE ✱ ✱

WWW.CHAPOUTIER.COM

Michel Chapoutier has managed that rare trick of creating a wide-ranging, successful wine business while remaining true to his values of terroir-driven wines that are respectful of the environment. The entire range includes certified organic and biodynamic wines, right down to the brilliant-value Côtes de Rhône and Beaujolais. His Ermitage wines are grown on the famous Hermitage hill, the acknowledged birthplace of the Syrah grape, and Le Pavillon is from low-yielding Syrah vines aged between 90 and 100 years old. One of my hero wines, the tobacco notes are clear, tight and smoky right now, but they mellow over time. The liquorice, wild cherries and fennel strike clearly, all wrapped up in velvety rich tannins. Garlic, thyme and rosemary are called for here – pick your favourite accompanying meat, fish or vegetable.

MAS CHAMPART CLOS DE LA SIMONETTE SAINT-CHINIAN 2014, FRANCE

BRAMEFAN, 34360 SAINT-CHINIAN

An increasing buzz surrounds the wines from this estate in the foothills of the Cévennes. Owned by Isabelle and Mathieu Champart, who moved to the Languedoc from Paris, the Clos de la Simonette bottling is a blend of Grenache (20%) grown on limestone soils and Mouvèdre (65%) and Carignan (15%) grown on clay, all aged for 18 months in large oak vats. Tight and young right now, it is a gorgeous peaty, smoky, graphite whirl of black fruits and dark chocolate. Gourmet, tightly structured, clove spiced. Pair it with something equally robust – I'm imagining a wild boar stew.

The Guibert Brothers
Mas de Daumas Gassac

HAUTE VALLÉE DU GASSAC, 34150 ANIANE, FRANCE
WWW.DAUMAS-GASSAC.COM

Impossible to select just one of these brothers. They all have a role to play in the running of the iconic Mas de Daumas Gassac. Gaël and Romain live and work on the Languedoc property, with Gaël taking lead as viticulturalist overseeing what has been an organic wine estate since it was first planted in 1974. Samuel is the winemaker, and splits his time between the Mas and San Francisco with his American wife, while their brother Basile was the most recent to join in 2013.

They are following in the footsteps of one of the true legends of French winemaking – their father Aimé Guibert. Protestant in both upbringing and outlook, Aimé was an artist in his approach to winemaking but also a tireless worker who has passed that trait on to his children (there are nine of them in total).

'When my parents arrived at this run-down spot, with a mill and a river running through the Gassac Valley, they fell in love with it immediately and began its restoration,' says Samuel. 'My mother looked at it as her garden. They drank the water from the springs and ate the vegetables from the garden. The idea of using chemicals was unthinkable, and she always said to dad that if he cut down one tree she would divorce him!'

> 'We all grew up here and Mas de Daumas Gassac runs in our blood. Our parents did all of this for us, and we in turn do it for our children.'
> *Samuel Guibert*

A perfect reflection of one of their father's favourite quotations. 'You can give only two things to your children: roots and wings.'

MAS DE DAUMAS GASSAC IGP SAINT-GUILHEM-LE-DÉSERT CITÉ D'ANIANE 2010, FRANCE *

This was a wild valley before it was planted with 80% Cabernet Sauvignon and 20% 'rare grapes' (including Tannat, Saperavi, Bastardo) by Amié and Veronique Guibert in the 70s. Now run by their sons, even cellarmaster Philippe Michel is basically part of the family – Samuel's childhood friend and best man. Mas de Daumas Gassac needs age – tasting back to 1985 was a delight. The 2010 is still young, but the fruit manages to be gorgeously warm and welcoming, especially after an hour in a carafe. It gets richer and deeper with every minute. Full of vigour, strong dark-fruit flavours, liquorice, black truffles, rich damson plums. No wonder it's called the *grand cru* of the Languedoc.

'This wine smells and taste of the French garrigue*, so serve with a roasted shoulder of Iberico pork laced with plenty of wild thyme and a rich Pommes Lyonnaise of sliced potatoes braised in stock and flavoured with chive flowers.'*
Ronan Sayburn MS, 67 Pall Mall, London

CAVALLOTTO BARBERA D'ALBA SUPERIORE BRICCO BOSCHIS VIGNA DEL CUCULO 2013, ITALY

WWW.CAVALLOTTO.COM/EN

This estate pioneered organics in Barolo back in the 60s, helped by their amphitheatre of vines that meant they could control their own environment. At a time when everyone was using chemicals, they began to work with researchers at the University of Agronomic Sciences in Turin, the Institute for Experimental Viticulture in Asti and several others in an attempt to reset the way they made wine. In 1976 they completely stopped all use of synthetic fertilizers, and more recently have severely reduced even the use of copper in the vineyards, replacing it with essential oils such as sage, ivy, aloe vera, yucca and various species of algae. Today run by Alfio, Giuseppe and Laura Cavallotto, this is really a property to watch. This Barbera d'Alba made using the Barbera grape instead of the region's more prestigious Nebbiolo still showcases the brilliance of Piedmont red wines, and is utterly delicious. Firm but beautifully integrated, smoky-edged tannins, it needs a good half an hour after opening to really loosen up and show off its raspberry and spiced damson fruits. We had it with a homemade lasagne – not very original perhaps, but wow it was good.

FATALONE GIOIA DEL COLLE PRIMITIVO RISERVA 2011, ITALY

WWW.FATALONE.IT

We are in the sun-baked heel of Italy here and this stunning winery is set on a rocky hilltop right in the middle of Puglia, equidistant from the Adriatic and the Ionian Seas. This Riserva is made from Primitivo, a grape that was born and that thrives here (you may also know it as its close cousin Zinfandel in California), and the intense sunshine translates into not only high alcohols – this might be the highest in the book for the still wines – but also rich, beautiful flavours that glide over your tongue. The Petrera family was the first to bottle the Gioia del Colle Primitivo as a single grape variety back in 1987 and Pasquale is a great defender of both this and the local white Greco grape. The winery is 100% sustainable – all electricity is produced by solar panels and the grapes are dry-farmed without irrigation. In the glass, you'll find the wine is rich, dark and almost menacingly spicy on the attack, but then softens and welcomes you to its heart of sour cherry, mulberry fruits and chocolate. Food-matching, they suggest game or salami, which sounds about right for such a powerful wine. One to crack open in winter when you want to fantasize about the heady, sultry days of a southern Italian summer.

FATTORIA LA MAGIA DOCG BRUNELLO DI MONTALCINO 2011, ITALY

WWW.FATTORIALAMAGIA.IT

One of original organic producers of Brunello, the Schwarz family has been farming this way since the 80s. This wine has the rich, yet soft tannins that you expect from Brunello, and the power and gentle roasting of the Tuscan sun. Don't expect the flashiest of Brunellos, this is more about precision and well-placed fine tannin, with unshowy but deeply satisfying rich-red-cherry flavours from the 100% Sangiovese grapes. An undercurrent of oregano spice, balsamic and the soft brush of *garrigue* places you slap-bang in Italy. Begs for a good ragu sauce.

FONTODI FLACIANELLO DELLA PIEVE 2006, ITALY

WWW.FONTODI.COM

This is one of Chianti Classico's most impressive properties, planted to 90% Sangiovese. With a bit of bottle age, this wine is heart-stopping. The Sangiovese grows in schist soils, and 2006 was outstanding, with highly aromatic and intense wines. Elegant, gentle spice makes you re-think the accepted idea that all Tuscan wines have to be big and burly. Giovanni Manetti not only converted his own estate to organics, but managed to convince most of his neighbours to do the same. Crostini with red peppers and rubbed down with lashings of garlic is delicious.

Above: Well-ripened Sangiovese grapes at Fontodi.

GULFI NEROBUFALEFFJ 2011 IGT SICILIA ROSSI, ITALY

WWW.GULFI.IT

Where should I start with what I like about Gulfi? The wines are unpretentious, good value and packed with personality. They never irrigate even in the Sicilian hot summers, they use local native grapes and keep sulphur additions to an absolute minimum. This Nero d'Avola is intensely herbal with intense blackberry and raspberry fruits, you can almost feel the pips they are so beautifully ripe, yet it is all carried out with an amazing lightness of touch for such a big wine. Just rustic enough with a lovely weight in the mid-palate, it opens up beautifully and pretty much demands that you stop and pour a glass for a friend. *Spaghetti alla puttanesca* would be a great match.

I VIGNERI VINUPETRA DOC ETNA ROSSO 2014, ITALY * *

WWW.IVIGNERI.IT

Oenologist Salvo Foto deserves every bit of praise that comes his way for this stunning wine. Part of a project that he founded between the local grape-growers of Etna, Vinupetra comes from vines located at 700m (2296ft) in altitude on the north side of the volcano where winters get cold and summers extremely hot. Many are more than 100 years old, giving dark black fruits, spicy wild sage and mint, with an almost bitter-amaretto biscuit aspect. You can practically taste the volcanic soils. This is gorgeous, with brilliant freshness running through it. Awesome with salamis, Parmesan, capers and breadsticks.

MATTEO CORREGGIA, RÒCHE D'AMPSÈJ DOCG ROERO RISERVA 2011, ITALY * * *

WWW.MATTEOCORREGGIA.COM

Almost 15 years on, incredibly, since the brilliant Matteo Correggia passed away; his family – wife Ornella and children Giovanni and Brigitte – has continued to make this one of the most exceptional wines in Piedmont. I first heard about them when I wandered in to a wine shop in Barolo to ask which producers I absolutely shouldn't miss while in the area. This was the first name suggested, and I have been grateful ever since. This particular wine is 100% Nebbiolo, grown at 300m (984ft) altitude on steep slopes of sand and silt underpinned by a powerful clay. It is a meaty, broad-shouldered, powerful affair, full of the nobility of Nebbiolo with a swagger of its own that more than demonstrates Roero can give Barolo a run for its money in the right hands. Look for firm tannins, textured red fruits and an undercurrent of fragrant spices. A pumpkin or squash risotto is a great choice here.

Jan Hendrick Urbach
Pian dell'Orino

LOCALITA' PIANDELLORINO, 189, 53024,
MONTALCINO, SI, ITALY
WWW.PIANDELLORINO.COM

Italian-German couple Jan and Caroline are not from Tuscany, but they have perfectly shaped their corner of it in their image. Jan grew up in the German countryside, and went to an agricultural high school. He wanted to study landscape architecture but when he didn't get the grades, turned instead to an apprenticeship in a winery.

'I was always interested in respecting nature, so organics for me was clearly easy to understand. My grandfather owned a winery, and I was always interested in landscape, in the English idea of copying nature in gardening. I love visiting places like Winston Churchill's garden at Chartwell in Kent. I earned my money as a student by selling bulbs in a local market. But when I started to work full-time in vineyards, I realized that things in reality didn't always respond successfully to organics, and I started to think there had to be something more.'

'I became interested in the rhythms of nature, about how outside influences help the plant react to things, and how the plant hormones are affected.'
Jan Hendrick Urbach

'When I read Steiner I found that he had a lot of answers to the questions that I had already been asking, and his suggestions were very interesting to me as a way to respect and work with the vines''

'I see biodynamics as preventative, not a cure, unlike organics. And increasingly I see how a vintage is not an entity in itself, but is the result of the previous vintage also. So much information is stored in the vine – the buds for the new season, even the grapes, much of how they behave has already been decided. So now we try to influence the plant positively for the year ahead as well as season that we are in. It is a constant challenge, and it is hard not to hold tension throughout the season, but there is a great satisfaction of thinking 'I did the very best that I could for these vines.'

PIAN DELL'ORINO VIGNETI DEL VERSANTE BRUNELLO DI MONTALCINO 2011, ITALY

Caroline Pobitzer and Jan Hendrick Erbach's 5.7-ha estate is right next door to the famous Biondi Santi, and is increasingly making all the right noises of its own. The wines come from four different vineyards, all planted to Sangiovese, all small-sized and biodynamic. This is big, bold, tight Brunello, with endless layers of black fruits given just the right amount of support by pliable tannins and fine acidity. It is firm, textured, with floral edges that flesh out the classicism of Brunello. Even at six years old the tannins are chewy, and this could age easily for another decade or more. I had this over a long supper of wild boar linguine, and it kept on subtly changing in the glass, offering different flavours, by turns wild herbs, liquorice, spice and rich fruits.

POGGERINO CHIANTI CLASSICO RISERVA BUGIALLA 2012, ITALY

WWW.POGGERINO-CHIANTI-ITALY.COM

This is a small family run winery in Chianti Classico with 11 ha of vines, plus olive groves and woods. Savoury plums and cranberries abound in the glass, this is a beautifully balanced Sangiovese, with just the right amount of earthy, chewy tannins and softly brushed leather to feel that the fruit has a partner cheering it on without it being even the slightest bit intrusive. Fermented in stainless-steel tanks then aged in large Slovenain oak casks for 18 months and held for 12 months before release. Sun-dried tomatoes, olive paste and fresh *burrata* – delicious.

PORTA DEL VENTO ISHAC NERO D'AVOLA IGT SICILIA 2015, ITALY

WWW.PORTADELVENTO.IT

A Nero d'Avola from the Valdibella district of Palermo, Sicily, from vineyards set at 600m (1969ft) in altitude. Marco Sferlazzo is the winemaker, working with the classic Nero d'Avola grape that is packed with sour but extremely juicy cherries together with grilled cinnamon and black pepper spices. Long, slow fermentation on its skins in open vats, then aged for a year in large oak casks. The wine has fresh acidity, especially for a wine from Sicily, and I absolutely loved the match with a chargrilled tuna steak.

SAN POLINO BRUNELLO DU MONTALCINO RISERVA 2007, ITALY * * * *

WWW.SANPOLINO.IT

This wine transports you to Tuscany, with silky tannins, liquorice and plum notes alongside an intense herbal concentration of rosemary and scrubland. It is full of impact but not so punchy that you can't just sit back and enjoy the ride. This producer is known for its long-lived wines, which is why I have chosen a 2007, you get the first hint of tertiary flavours, with truffles dancing alongside the tighter olive notes. The estate has vines, olive trees and extensive forests, and works only with Sangiovese. They are also involved in a fascinating biodiversity-mapping project in the Brazilian Amazon, and they practice permaculture back in Italy. Hugely food friendly as is all Italian wine, but from experience I can tell you this is astonishing with a spinach cannelloni.

Giuseppe Maria Sesti
Sesti Wine

CASTELLO DI ARGIANO, 53024 MONTALCINO, SIENA, ITALY
WWW.SESTIWINE.COM

The glasses might make Giuseppe Sesti one of the most recognizable winemakers in Italy, but it is his wide-ranging intelligence and the beauty of his wines that make him one of the most sought after. It's almost impossible to believe that Giuseppe only began making wine in 1991, such is the reputation today of this Montalcino estate. But then this is a man who as a student would spend hours in St Mark's Biblioteca Nazionale Marciana library in Venice, reading 'hundreds if not thousands of books' because he was friends with the librarian who would let him in and then leave him in peace.

'This was where I first started seriously learning about ancient calendars and astronomy, although at the same time I also used to go to a farm in Valdobbiadene for the grape harvest every autumn where I met Silvano, the grandfather of the family who could hardly write a word and yet knew every aspect of farming and viticulture, how to make farm tools, how to assist animals giving birth and many other ancient crafts. It was my first encounter with the power of oral tradition.'

Today it is Giuseppe who has this knowledge to pass on. He has written several books on the subject of astronomy, and has integrated the belief in the power of celestial spheres into his winemaking.

'I am sitting on the best possible land for vineyards, surrounded by a green lung of oxygen, with the right orientation and exposure for the Sangiovese grape to ripen perfectly. It is a grape that needs the right vineyards in the right place, and when harnessed to the power of natural cycles can be magical.'
Giuseppe Maria Sesti

SESTI CASTELLO DI ARGIANO BRUNELLO DI MONTALCINO 2012, ITALY

This just outclasses pretty much any Brunello you care to mention, and is dripping with wild strawberries and blackberries, sweet balsamic, tar and freshly cut herbs. It is a classic, almost understated compared to some, and yet one that packs a punch of flavour and whose memory stays with you. To give you some idea of the scale of biodiversity here, there are 102 ha of land, with just 9 ha of vines and the rest olive groves, pasture and forest. The wines are a fitting testament to this rich diversity. They have chosen to not be certified in organics or biodynamics, but no chemicals are used in the vines, and the cellar practices are minimum intervention, with 36 months in neutral (read no new) oak and 12 months in bottle before release. So what to choose to eat with this? Giuseppe's daughter Elisa says, 'In Tuscany wine and olive oil are considered food, and should always be on the plate in balance with the freshest food of the season.'

Opposite: The stunning Sesti estate in Brunello di Montalcino.

STELLA DI CAMPALTO ROSSO DI MONTALCINO 2012, ITALY

WWW.STELLADICAMPALTO.IT/EN

Stella di Campalto has been called the 'accidental winemaker', because the vines at the San Guiseppe estate came to her via a wedding gift. Originally founded in 1910 but abandoned in 1940 until her family arrived in 1992 – the vineyard eventually passed on to Stella. She studied for one month in Bordeaux, but says 'the real learning was from the land, getting to know Sangiovese' and advice from neighbouring winemakers. It was aged first in oak casks for 19 months then in bottle for 21 months before release. Juicy, intense, a wonderful mix of spicy cloves, white pepper, soft tobacco, rich redcurrant jelly and crushed cherries. I had it with a wild mushroom risotto – heavenly.

TENUTA DI CAPEZZANA, VILLA CAPEZZANA DOCG CARMIGNANO 2011, ITALY

WWW.CAPEZZANA.IT

Carmignano was the first wine region that I ever got excited about, but I have lost touch with it in recent years, so this was a wonderful rediscovery. Capezzana is the biggest estate here, run by Vittorio Contini Bonacossi. It was certified organic in 2015 but has been farming this way since 2009 – they don't even use organically approved insecticides and prefer to encourage natural predators. Carmignano wines were essentially Super Tuscans before the term was even coined, because they have always blended Cabernet Sauvigon and Cabernet Franc in with their local grapes. This is a blend of 80% Sangiovese and 20% Cabernet Sauvignon, and is rich, deep, spicy, full of dark plums, olive paste and rich chocolate. For a food match, you can't go wrong with a slow-braised piece of beef.

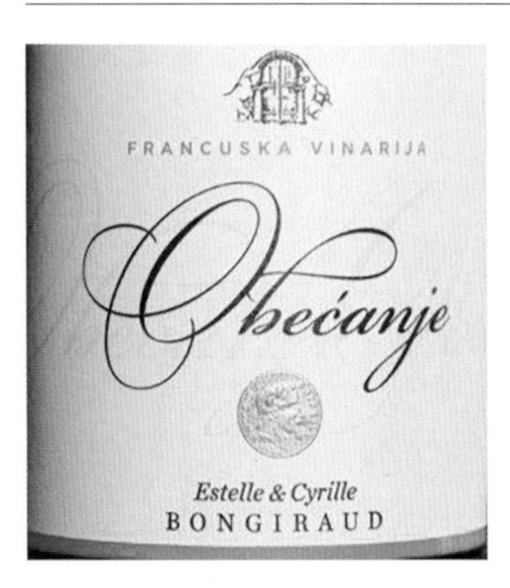

BONGIRAUD FRANCUSKA VINARIJA OBECANJE 2013, SERBIA

WWW.LESBONGIRAUD.COM

Husband and wife Cyrille and Estelle Bongiraud are responsible for rescuing this almost-lost piece of viticultural heritage in Serbia, and have helped revive the artisan traditions of a number of small villages in the region. Obecanje means 'promise', and this wine, from Gamay à Petit Grains, stands testament to the history of Serbia's winemaking, as the grape was gifted in the 19th century by French winemakers. It needs time to open, but then displays a striking purity of fruit, clear white pepper, mint and fresh herbs on the nose, followed by lush blueberries and bilberry fruit, with subtle acidity. Easy drinking and refreshing, surprisingly delicate for its alcohol level. No oak, and low sulphur, all vivacity. Get some grilled lamb going to partner with this.

CHATEAU MUSAR, BEKAA VALLEY 1999, LEBANON ✱ ✱ ✱

WWW.CHATEAUMUSAR.COM

The legendary Chateau Musar, founded in 1930 by Gaston Hochar, was the first estate to be certified organic in the Lebanon, back in 2005. It was another first in the history of a property that has always inspired fascination and devotion by equal measure. Gaston had moments of flirting with modernity because he was close to many of the big estates in Bordeaux that began implanting modern winemaking practices in the 50s. It was his son, Serge Hochar who died aged 75 in 2016, who pushed back against all that, wanting to keep the winemaking as traditional and non-interventionist as possible. 'It was his philosophy, his approach to life,' says his son Marc today. 'So Musar escaped the winds of modern winemaking.' We can all be grateful that he stuck to the original personality of this wine. The first thing that strikes you on opening the bottle is the fragrance. It's exotic and enticing, both floral and spicy, and curls out of the bottle even before being poured into a glass. This is almost 20 years old and there are definitely tertiary aromas, even a touch of dried figs and dates. But that's just the first act. This is a wine where the real action starts in the mid-palate. It's savoury and spicy, and you feel the acidity and tannins still cushioning the fruit. Musar is a wine to drink with serious bottle ageing – this is the vintage that they recommend to be drinking today, and tell me it is still available to buy because they deliberately keep wine back to encourage later consumption. Go Middle Eastern with this: slow-cooked lamb, tabouleh, the ever-brilliant Fattoush salad of pitta, tomatoes, cucumber, olive oil and plenty of fresh mint.

JAKELI WINES KHASHMI SAPERAVI 2011, GEORGIA

WWW.JAKELI-WINES.GE

A small winery, certified organic since 2009, located on the foothills of the Tsiv Gombori mountain range. This estate is run by brothers Zaza and Malkhaz Jakeili who are self-taught and only got started in 2001. I'm not going to pretend that it is particularly easy to find this wine, but it's worth it if you can. Saperavi is inky and rich in colour. The thing that surprised me the most was how fruity it was – it is deeply coloured but has clear plum and pomegranate flavours and relatively low tannins. The acidity is quite high, but it's the spicy edge that tells you that this is not your everyday wine. Fermented with wild yeasts, it is aged for 30 months without oak and bottled unfined and unfiltered. I enjoyed this with spicy meatballs in a rich tomato sauce.

Maxence Dulou
Ao Yun

SHANGRI-LA, YUNNAN, CHINA
WWW.LVMH.FR/MEDIA/AO-YUN

You need to be patient, stubborn, resourceful and self-contained to make wine up in the Himalaya mountains at 2600m (8350ft) where electricity regularly cuts out and the drive is a good three hours from the nearest town of any size. And that's before taking into account the winemaking skills necessary to make a world-class Cabernet for the demanding bosses at Moët Hennessy.

His success is probably why Bordeaux-born Maxence Dulou, who heads up the project, is known as the magician. He can fix the generator, repair the boiler and pour you a good glass of wine at the same time. And he is smart enough to know that he knows nothing compared to the people who have lived in the region all their lives.

'We work with 120 farming families who have worked organically by hand for centuries, they understand the plants very well and our quality objectives would not be possible without their knowledge. They built dozens of terraces long ago and live autonomously by creating a virtuous circle from plant to animal – using weeds, leaves and shoots from the vines for animal feed for example. Between each terrace there is a slope planted with selected wild plants that will feed their black pigs and yaks, or edible vegetables for the farmers and their families. Walnut trees are planted at regular intervals providing shade for all. But we have to constantly balance our requirements with their own cultural needs of looking after their plantations, such as collecting the famous culinary and medicinal mushrooms in spring and summer.'

'It's a constant challenge, and one that I will always be grateful to have lived.'
Maxence Dulou

AO YUN 2013, CHINA

Award for the most unusual vineyard in this book just might have to go to this one – the wine is grown in small plots across four villages in the Himalaya mountains in the Yunnan region of China just a few miles from the border with Tibet. I've visited these vineyards and this is handmade wine at its purest. No machinery could ever make it down the tiny dust tracks that lead to the vines, and its remote spot means self-sufficiency is the order of the day. Owned by LVMH, Ao Yun (meaning 'above the clouds') is a blend of 90% Cabernet Sauvignon and 10% Cabernet Franc. This is the first vintage: bold and dramatic, full of bilberry and graphite notes; aged in 100% new oak it has a polished finish and big tannins, achieved with low levels of sulphur because the altitude offers natural protection against oxidation during winemaking. This could stand up beautifully to braised short ribs, or braised red cabbage.

Toby Bekkers

Bekkers Wine

212-220 SEAVIEW ROAD, MCLAREN VALE, 5171 SA,
AUSTRALIA
WWW.BEKKERSWINE.COM

A huge influence in biodynamic farming in Australia, Toby Bekkers consults or is good friends to half of the biodynamic farmers in the country as far as I can tell. He was also incredibly kind and helpful to me when researching this book.

A trained viticulturalist, he worked with David Paxton, the original biodynamic producer in McLaren Vale, and helped convert his vineyards before heading off to spend a year in France (where his wife Emmanuelle is from) before returning home in 2011 and striking out on his own.

Besides being a brilliant winemaker, he has that beautifully Australian trait of practicality in spades, telling me bluntly, ' I've a fair bit of history with organics and biodynamics. I now focus mainly on the Bekkers wines but also act as a consultant viticulturist around Australia, mainly in McLaren Vale and Margaret River. I've arrived at a point where we build tailored farming systems to suit particular businesses. This means some adopt certification while others don't. Some choose to use a little from both conventional and organic systems. The Bekkers wines are predominantly sourced from vineyards managed under organic and BD regimes (including the one in this book). I also source some conventionally farmed fruit when it's in a superior site with a view to working with the owners to transition to softer practice.'

> 'I guess I'm trying to outline that I'm not a fundamentalist for any one system and that our wines don't have organic certification, nor would we seek it.'
> *Toby Bekkers*

BEKKERS WINE SYRAH MCLAREN VALE 2015, AUSTRALIA

This wine has instant impact: rich damson in colour, with intense black spice, it balances the punch and intensity of an Australian Shiraz with the elegance of the same grape from the Northern Rhone. It seems to elongate in your mouth, with silky tannins and dark fruits backed up by chocolate notes and crunchy redcurrants. Delicious, it deepens after 15 minutes of opening, a brilliant example of why Toby Bekkers is such a sought-after consultant for organic and biodynamic farming. This is not certified, 'and nor do I want it to be,' he says. He leaves 15% whole-bunches in the vinification, with the rest destemmed but not crushed, with natural yeasts and extremely gentle extraction, all aged in French oak barrels (40% new) and given minimal sulphur at bottling. I had this with Chinese duck pancakes and it was a great match.

D'ARENBERG THE DERELICT VINEYARD GRENACHE MCLAREN VALE 2012, AUSTRALIA

WWW.DARENBERG.COM.AU

The name of this wine comes from the vine-pull scheme instigated by the Australian government in the 80s, when many old vineyards were in danger of being lost. Chester d'Arenberg and his father D'arry came to their rescue, buying up many overgrown or abandoned bush vines. 'They see themselves as the custodians of Grenache in Victoria,' is how winemaker Toby Porter puts it. This is a delicious and brilliant value wine. You can go right up to more expensive single-vineyard bottlings with d'Arenberg, but I love this one: 100% Grenache from Blunt Hills, where you find Grenache with aromatic floral and feminine notes. Some lovely blueberry fruits, earthiness and great natural acidity make it a great food wine – try smoked food like duck, ribs or sausages. If you have the chance, go cellar door here, and see the huge d'Arenberg cube that opened in May 2017. A kind of wine museum, art gallery, wine centre, shop and restaurant – all in a building shaped basically like an overgrown Rubik's cube.

PAXTON VINEYARDS QUANDONG FARM SINGLE-VINEYARD SHIRAZ MCLAREN VALE 2015, AUSTRALIA ✱ ✱

WWW.PAXTONVINEYARDS.COM

The original biodynamic grower in the McLaren Vale, David Paxton has been working his family estate since 1979, and has done huge amounts to convince the wider wine community of the benefits of this approach to farming. Any remaining sceptics should taste this wine. Fragrant raspberry and rosemary on the nose, with olive tapenade gliding into the fruit on the palate. You can feel the Australian sunshine, but it is beautifully balanced with the elegance and finesse that a Shiraz can deliver in the right hands. With soft tannins, this is fragrant and vibrant while still being big and bold – gorgeous and clearly able to improve in bottle for years to come. Oh, and a good story about why it's called Quandong Farm: when David bought the property, he knew that he'd get the land cheaper for normal farming than for vineyards, so told the agent that he was looking to plant Quandong trees. Who says environmentalists can't be smart businessmen too? Food-wise, oven-roasted root vegetables would be wonderful.

Peter Fraser

Yangarra Estate

809 MCLAREN FLAT RD, KANGARILLA SA 5157, AUSTRALIA
WWW.YANGARRA.COM

Down to earth, practical, totally straight-talking, it's no surprise that when Peter Fraser first decided to go organic, he didn't want the 'added pressure' of certification.

> 'But ultimately we decided that we wanted to distinguish ourselves from the many wineries who dabble in this stuff.'
> *Peter Fraser*

His reasons for appreciating this kind of work are also refreshingly practical. 'Organics and biodynamics are about all the little things that help you be more of a farmer, and that in turn helps you grow better grapes. And once you are getting there in the fields, you can't help but extend it into the winery. You start thinking instead of relying on tartaric acid, why don't I just pick a little bit earlier? And then how can I make the wine more reflective of the vineyard, and still make it taste good?' Peter grew up on a farm (where they worked conventionally), so a lot of this stuff made sense from the start, but it is still impressive to be working over 100 ha of vines biodynamically.

'All of this just adds another level of integrity to what we do,' he says. 'I like to keep things practical. All winemakers talk about creating a sense of place in their wine, but how can you be genuinely delivering that sense of place if you have blocked the pathways for a conversation between the soil and the plant? So for me it's about encouraging as much microflora as possible and invigorating the soil. In warmer regions like ours, the risk if you're not paying attention is that jammy overripe flavours can creep in'. Then he pauses. 'But some people like that. It's horses for courses.'

YANGARRA ESTATE VINEYARD OLD-VINE GRENACHE MCLAREN VALE 2014, AUSTRALIA *

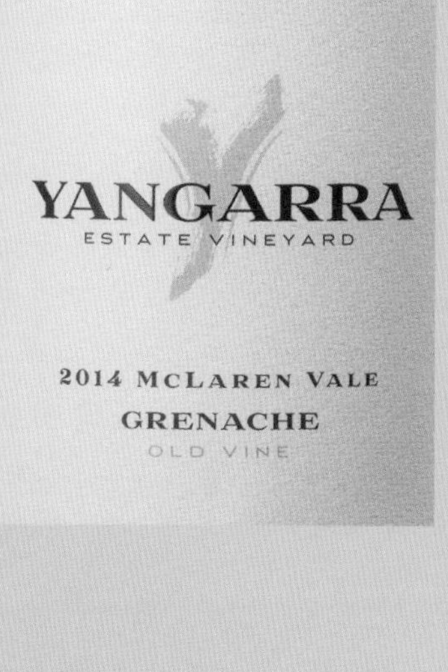

Bought by Jess Jackson and his wife Barbara Banke in 2000, and run by winemaker Peter Fraser and viticulturalist Michael Lane, Yangarra is a consistent force for excellence in McLaren Vale. Fully certified in biodynamics and organics across vineyards and winery since 2013, this is a serious dry-farmed estate with over 700 sheep for pasture management. An intense wine, with deeply spiced notes and a loganberry, slightly sweet edge; liquorice and smoke ripple through and there is a beautiful balance that brings lift, relief, freshness. It is open fermented with natural yeasts, bottled unfiltered and unfined.

'Bursting with vibrant red and blackberry fruit, intense mint and rosemary hint lifts the aroma. De-bone a lamb leg, grill outdoors on a barbecue, studded with lots of rosemary sprigs or served with a fresh mint relish.'
Ronan Sayburn MS, 67 Pall Mall, London

05

Digestifs

Robert Bath MS

Master Sommelier at the Culinary Institute of America, California

I was waiting on tables in fine wine restaurants at the age of 19 and writing wine lists by the age of 25, so food and wine have been a life-long love affair. I was lucky enough start at a time when I was introduced to the great German sweet wines during some of their greatest ever vintages – the 1971, 1975, 1976 Trockenbeerenauslese (TBA) and Eiswein from producers like JJ Prüm, Weingut Robert Weil and Egon Müller. I also fell in love with this style of wine through discovering the late-harvest Zinfandels from the 60s and 70s from producers such as Ridge and Caymus.

Together, these wines helped me understand how the flavours of wine and food are so much part of the same experience because the best dessert wines are almost a meal in themselves.

The attraction of sweet wines for me is partly about the idea of chasing something rare, it's like being given a gift because certain methods of production such as botrytis (noble rot) mean that they don't happen every year, or result only in tiny quantities of wine. They are artisan by their very nature, which makes them a natural fit for producers working organically and biodynamically. At the same time everything is magnified in sweet wines – concentration, flavours, length on the palate, ageing ability are all heightened.

Their rarity and artisan character are no doubt why dessert wines once commanded the highest prices of any wines in the world.

More rare and certainly more consistent than dry wines of their era, kings and queens in the 18th and 19th century appreciated these wines with a variety of foods other than dessert. Indeed, dessert wines still have a place early in the meal as Sauternes and foie gras so aptly demonstrate. And if our global palates have drifted more towards dry wines, the result is that many dessert wines are exceptional values in today's marketplace.

One of the challenges with dessert wines is that they come at the end of the meal when previous wines have had their moment and your palate is frequently less fresh and ambitious.

But the biggest challenge with dessert and dessert wines is that in many cases either can be quite extraordinarily independent of each other. The wines can be so grand and complex in flavour (as can be the desserts themselves) that putting them together can seem to lessen the overall effect rather than adding to it.

To get the best out of these pairings, there are some basic groupings and rules to follow. Dessert wines should initially be grouped by sweetness, alcohol, weight and acidity. Think about the level of these three components in the dessert and the wine.

The level of sweetness in both the wine and the dessert is a key point in pairing the two. The classic adage that the wine must be sweeter than the dessert is key.

If the wine is not sweeter than the dessert the wine will taste flat and too dry. You need to remember that sweetness in wine adds to the mouthfeel but can also make it taste heavier, so consider the weight and alcohol levels before selecting the food. A sweeter, high-alcohol dessert wine needs something heavier or denser to pair with – think a combination of chocolate with Port or a French Vin Doux Naturel. A delicate crème caramel loves a lighter Vouvray Doux.

Acidity in the wine refreshes the palate and can be helpful with making richer desserts taste lighter and helping acidic desserts taste richer.

Other factors to consider are effervescence, botrytis character, barrel treatment and age. These can be excellent complementary flavours like honey in the wine and honey in the dessert or the vanillin of an oaky wine and vanilla in the dessert. And certainly the butterscotch/toffee aspect of older or solera-based wines love to see that same component in the dessert they are paired with.

Effervescence in a dessert wine will never let you down. Moscato d'Asti may be the most versatile dessert-pairing wine on the planet. Its lightness, relative sweetness, acidity and that gentle fizziness makes it work with just about everything including fruit-based desserts, which can be very challenging to pair wines with.

Robert's Wine Picks

BERA VITTORIO E FIGLI CANELLI DOCG MOSCATO D'ASTI, PIEDMONT, ITALY

CHÂTEAU GUIRAUD, SAUTERNES, FRANCE

FROG'S LEAP FROGENBEEREN-AUSLESE, NAPA, CALIFORNIA, USA

JJ PRÜM RIESLING TBA (1971), MOSEL, GERMANY

WEINGUT BRÜNDLMAYER GRÜNER VELTINER LAMM TBA, KAMPTAL, AUSTRIA

SOUTHBROOK NIAGARA PENINSULA VIDAL ICEWINE 2014, CANADA

WWW.SOUTHBROOK.COM

The first vineyard in Canada to receive Demeter certification. Its location by Niagara-on-the-Lake gives it those helpfully bitter winters where temperatures drop down fast (you need -8˚C/17.6˚F to make icewine) and allow the grapes to freeze on the vine. This has 195g/l residual sugar and is made from 100% Vidal, which is picked during the coldest part of the night and pressed first thing the next morning. It delivers an almost savoury yet luscious blend of dried orange peel, peach, salty caramel and searing key-lime-pie edging. Personally I think icewines might just be the most interesting of all sweet styles to pair with desserts – even cheesecake works brilliantly.

red tail ridge
Block 907
2013 RIESLING
FINGER LAKES

RED TAIL RIDGE WINERY RIESLING BLOCK 907 FINGER LAKES 2013, USA

WWW.REDTAILRIDGEWINERY.COM

The Finger Lakes are producing more and more great Rieslings, and I was thrilled to be able to include this one in the book. Red Tail is found on the western shore of Seneca Lake, and has been really pushing the boundaries in terms of sustainable winemaking. Although they are not certified organic, they use integrated pest management to avoid insecticides and use no herbicides. This particular wine comes from botrytis grapes given a long, slow fermentation with natural yeasts. It has 82.52g/l residual sugar, and a beautifully bracing pH3.32, which explains why it just zings right off the glass.

'This one knocked my socks off. High acidity and super light. Has a tension that refreshes the palate and makes your mouth water for more. Orange oil and honey are both evident, and man this is a great wine! Still young enough that it was bouncy and fresh but also pretty and elegant. You could sit around and enjoy this with apricot jam and sugar cookies all day long, but it would also love a stinky blue cheese.'

Jeff Harding, The Waverly Inn, New York

David Guimaraens
The Fladgate Partnership (Fonseca)

PO BOX 1311, EC SANTA MARINHA, 4401-501 VILA NOVA DE GAIA, PORTUGAL
WWW.FONSECA.PT

Though David's perfect English accent might not give away much about his origins, he grew up here, he lives and breathes Port, and is pretty much a local legend. He is championing organic Port, a relatively new development in the region. It was tough to find an organic Port to include in this book, even though Fonseca has several organically farmed vineyards.

'The challenge of organic Ports is not in fact running the vineyards but sourcing an organic spirit. It wasn't until I found an organic spirit in 2004 from one of our Spanish spirit suppliers that I was finally able to make this.'

'We started converting parts of our vineyards to organics back in 1992. There are two main challenges here, early in the season there is the risk of powdery mildew and the steep slopes make it extremely difficult to manually control weeds.'

You totally get why when you take a trip out to the stunning Douro Valley, where vines seem to cling to impossibly steep slopes on all sides, unless large terraces have been cut into the mountains.

'It's been a huge learning curve in terms of working in this way. There is so much manual work involved in mountain viticulture, so the respect for the workers is key. Today I believe that any professional farmer should have a plot of organics because it forces you to get close to the vineyards.'

'I grew up in a period of a tremendous desire to mechanize our valley, and I now view that as a mistake – we should look instead to simplify. A lot can be achieved by doing this, striving to overmechanize has led people to upset the balance.'
David Guimaraens

FONSECA PORTO TERRA PRIMA, PORTUGAL

Somewhat surprisingly this is the first 100% certified organic Port, and one of the few truly organic Ports in the Douro. Developed by David Guimaraens, the grapes are farmed at Fonseca's Quinta do Panascal (farmed organically since 1992). Aged for four years in neutral oak before bottling, expect a ruby-style Port, which means it's all about bright but extremely dark damson and kirsch cherry fruits, all wrapped up in spiced vanilla pod. With 96.8g/l residual sugar, this is made from Touriga Nactional, Touriga Franca, Tinta Barocca, Tinta Roriz, Tinta Cão and Tinta Amarela grapes. The classic pairing with Stilton is going to stand you in good stead here, as would a few small squares of an extremely dark chocolate, ideally a single plantation bar from an organic producer like Akesson's.

Above: The Sauternes region of Bordeaux is famed for its graceful 18th and 19th century châteaux.

MAS ESTELA GARNATXA DE L'EMPORDA SOLERA 1990, SPAIN

WWW.MASESTELA.COM

Owned by the Soto-Dalnau family, this Catalonian estate has been farmed organically since Diego and Nuria arrived in 1989 (they stumbled across its abandoned ruins while out hiking). Aged in a solera system with the oldest wine dating back to 1990, this is made from 100% Grenache, aged in large oak barrels. With 125g/l residual sugar, this is a beautifully sticky red wine, drunk best slightly chilled. Flavour-wise you get a mix of burnt caramel orange, smoky and sweet, bursting with late summer fruits. It would be a brilliant match to a treacle tart and a cup of espresso.

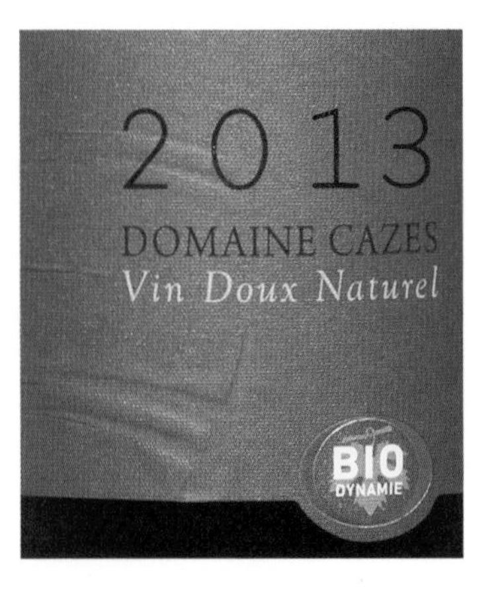

CAZES RIVESALTES GRENAT 2013, FRANCE

WWW.CAZES-RIVESALTES.COM

Cazes is the largest organic and biodynamic estate in France, covering 220 ha. You have a wide range of Vins Doux Naturels to choose from, including a large selection of old vintages, some dating back to the 30s, where just 500 bottles of each are released. I have chosen the crowd-pleasing Grenat bottling. It's a 100% Grenache Noir wine that comes from 30-year-old vines with extremely low yields, grown on schist and gravel soils. With around 90g/l residual sugar, you can expect kirsch cherry, overlaid with black olives, rich mocha chocolate and cigar box flavours. Lay this out at the end of a meal next to a round of black coffees, a plate of biscotti or a dark-chocolate mousse and away you go.

Bérénice Lurton

Château Climens

33720 BARSAC, FRANCE
WWW.CHATEAU-CLIMENS.FR

Part of the sprawling Lurton clan, Bérénice shows us every single vintage and just why Sauternes wines deserve to find their way back into the public consciousness. And it's hard not to see that the way she farms, which results in a natural sense of vitality and freshness, is a particularly brilliant way to showcase their charms.

Bérénice took over at the property (which itself dates back to 1547) from her father Lucien in 1992, and began farming biodydnamically in 2010, gaining certification in 2014.

'For a number of years I knew that farming traditionally – by which I mean using conventional treatments – was simply not sustainable over the long term, and yet I have never felt that organic farming is well suited to the oceanic climate of Bordeaux.'

'I wanted to do something to emphasize the personality and vitility found in the older vintages of Climens. In 2009 I met with Jean-Michel Comme in Pauillac and was bowled over by the holistic approach to winemaking at Pontet-Canet. It resonated so strongly with me, and Jean-Michel's wife Corinne Comme has been a great help to me and my technical director Frédéric Nivelle as we made the conversion to biodynamics. For both of us it has been a revelation, a wonderfully energizing way to expand our horizons and way of looking at the world.'

'Biodynamics is not easy, but I truly believe it offers the best way to rediscover the truth of our terroirs and our wines.'
Bérénice Lurton

CHÂTEAU CLIMENS GRAND CRU CLASSÉ BARSAC 2014, FRANCE

2014
Château Climens
1er Cru · Barsac
GRAND VIN DE SAUTERNES
Bérénice Lurton

There's a high-ceilinged loft at Château Climens where Bérénice Lurton and her team dry flowers and herbs for the biodynamic infusions. Hessian sacks are lined up, heaped with chamomile, nettle and other plants picked from the château gardens. It's peaceful up there, a perfect place to feel the benefits of this natural system of farming, which is not always easy for sweet wines made from noble rot. Climens is testament to its benefits. This is spun silk, precision-winemaking in the extreme, with shots of blood orange and quince dancing next to tight lime flavours and rich lemon curd. The heady aromatics of white truffles, apricot tart and lemon zest just curl right up out of the glass at you. This is 100% Sémillon with a knife-edge tension. My advice for food matching here is to keep things simple – slice a few perfectly-ripe apricots and peaches, sprinkle on a few toasted almonds and enjoy.

Xavier Planty

Château Guiraud

33210 SAUTERNES, FRANCE
WWW.CHATEAUGUIRAUD.FR

If you need imagination and determination to convert to truly sustainable winemaking in an area that positively encourages rot as Sauternes does, then Xavier Planty has it in spades.

Born in 1955 in Bordeaux, his first degree was in biology, with a focus on plant genetics, so it is fair to say that he has always been interested in the more natural side of science. He has run Château Guiraud since 1986 and been a co-owner along with three friends (Robert Peugeot, Olivier Bernard and Stephan von Neipperg) since 2006.

Things can get pretty traditional in this part of Bordeaux, but ever since I met Xavier 15 years ago he has been the source of countless innovations – from insect hotels in the vines to a selection programme for getting higher-quality Sémillon vines into the region's vineyards to hosting an annual Full Moon Party for the Chinese New Year. Oh, and not content with getting certified in organics, Guiraud has come up with the term Bio-Viticulture to reflect the work that they are doing to rebalance the entire ecosystem.

'Among the many benefits for the wine of the biodiversity introduced in the vineyard, we are now able to add estate-grown morel mushrooms to the table, as since going organic kilos and kilos of them grow every year!'
Xavier Planty

CHÂTEAU GUIRAUD PREMIER GRAND CRU CLASSÉ SAUTERNES 2011, FRANCE

An exceptionally good vintage for Bordeaux's sweet wines; with bottle age this wine is starting to show just how multi-faceted it is. A blend of 65% Sémillon and 35% Sauvignon Blanc, it showcases Sauternes' beauty: ginger, mango, papaya slide into caramelized baked apples and toasted vanilla; as that first explosion of fruit subsides, the more biting flavours of lemon, lime and mint kick in. Chicken and morel mushrooms could nicely accompany the truffle flavours in an older Sauternes.

'Seasonal peaches poached with thyme and honey, served with a gingerbread sponge, reduced poaching syrup and a sage ice cream.'
Henrik Dahl Jahnsen, Bølgen & Moi, Norway

CHÂTEAU TOUR DES GENDRES CUVÉE DE CONTI 2014, FRANCE

WWW.CHATEAUTOURDESGENDRES.COM

The de Conti family moved to France from Italy in 1925 and today are one of the go-to names in the Bergerac region. They have been farming organically since 1994, moving over entirely in 2005, and are increasingly hands-off in the cellar. This is an excellent-value, dry 100% Sémillon wine made with late-harvest raisined-on-the-vine grapes. It majors on exotic fruits, pear and quince, and has a lovely persistency. It's particularly great with cheese, especially if you load your plate up with an array of dates, quince and dried apricots alongside a crumbly blue. Or if you feel like getting stuck in to local French traditions, it would also pair beautifully with foie gras. Another unusual wine of the de Contis to try is their Pét-Nat from 100% Sauvignon Blanc vines, bottled with no added sulphur.

DOMAINE DES ENFANTS MAURY 2011, FRANCE

WWW.DOMAINE-DES-ENFANTS.COM

Luscious wines from Swiss-American couple Marcel Bühler and Carrie Sumner, who plough the venerable old vines (mainly 65 to 100 years old) by horse and weed by hand and picket. No chemicals, wild yeasts and minimum interventions at every stage in the cellar. Dark cherry, raspberries and lashing of rich chocolate. This is a wine that I love to take to friends' as a gift, usually pairing it with chocolate too, because it is indulgent, unusual and always an unbelievable crowd-pleaser to pull out at the end of an evening. Ecocert-certified.

DOMAINE DE L'ANCIENNE CURE L'ABBAYE, MONBAZILLAC 2015, FRANCE

WWW.DOMAINE-ANCIENNECURE.FR

It is hard to beat this wonderful sticky wine from Christian Roche, owner of this celebrated 47-ha certified-organic estate in the Bergerac region of Southwest France. Quince, mandarin, bitter oranges abound. Rich, tart and juicy all at the same time, you will be happy that the flavours just don't know when to quit. Drink now or put it away and bring it out in a few years time. A blend of Muscadelle and Sémillon.

'Supper zippy, amazing with a gluten-free almond flour cake with whipped mascarpone and candied orange peel.'

Jeff Harding, Sommelier, The Waverly Inn, New York

DOMAINE DU TRAGINER BANYULS GRAND CRU HORS D'AGE NV, FRANCE

WWW.TRAGINER.FR

Old school in the extreme, Jean-François Deu doesn't like to answer the telephone or email, and prefers to let his wine do the talking. But if you do get the chance, visit his wonderful vineyard in Banyuls-sur-Mer. Stunning steep-sloped terraced vineyards on schist soils overlook the sea and are worked by mule, with sheep grazing during the winter months. This utterly indulgent Banyuls is a blend of 65% Grenache, 25% Grenache Gris and 10% Carignan. Take your time with this wine, roll it around the glass and your mouth to get the full impact of the sweet but never cloying richness of chocolate, figs, coffee, burnt caramel and tobacco flavours. It is bottled unfiltered and unfined, with 93g/l residual sugar, after ten years in large oak casks. One of the very rare wines that can pair with chocolate, so make the most of it.

DOMAINE EUGENE MEYER RIESLING VENDANGES TARDRIVE 2011, FRANCE

WWW.EUGENE-MEYER.FR

This estate has the distinction of having been the first to receive official biodynamic certification in France, working in this way since 1969. Light and fresh in style, made from late-harvested grapes that reached natural over-ripeness with some noble rot forming but not fully shrivelled on the vine. Fermented from its natural yeasts, the final wine has 41g/l residual sugar and conveys a balance of tension, minerality and explosive apricot, white pepper, cinammon spice and ripe, grilled peaches. Riesling just does sweet wines so well, it never ceases to amaze me. You could happily drink this on its own, or with a plate of grilled almonds, or go full-on apple-tarte-tartin and vanilla cream.

DOMAINE HUET CLOS DU BOURG PREMIÈRE TRIE VOUVRAY MOËLLEUX 2005, FRANCE

WWW.DOMAINEHUET.COM

A legendary estate made famous under Gaston Huet, today owned by Sarah Hwang. Sadly long-term winemaker (and Gaston's son-in-law) Noël Pinguet also left a few years ago, but the wines remain among the greatest examples of Chenin Blanc in France and should rightly be celebrated as such. The grapes come from the famed 6-ha Clos du Bourg vineyard, with shallow clay-limestone soils. It is aged 50% in cement tanks, 50% in large oak vats, and the 81g/l residual sugar seems to melt into insignificance, leaving just a caress across your tongue. It is stunning now, but the white truffles and candied lemons will only increase over time. For food pairing, try a salty plate of oysters or caviar. Or save it for the cheese board.

DOMAINE MARCEL DEISS ALTENBERG DE BERGHEIM GRAND CRU 2011, FRANCE

WWW.MARCELDEISS.COM

Jean-Michel Deiss has made a career out of doing things his own way. This is a prime example, with a wine made from 13 different grape varieties including Riesling, Gewerztraminer and Pinot Gris that showcases the historic ways of planting in Alsace rather than the 100% varietal approach that has become standard here. His reasoning is that it's the best way to let the terroir speak louder than the grape, and certainly you stop thinking about variety when you are enjoying a glass of this. Your mind instead takes you to marmalade, roasted almonds, lychee, peaches and ginger, all intensified by sunshine on the fully south-facing slopes of Altenberg. It has the body to stand up to a variety of flavours, but would do best with a not-too-sweet dessert like crème caramel that could pick up on the rich grilled fruit flavours.

Above: Grass growing between rows is common in organic vineyards, as in Domaine Huet's Clos du Bourg.

Olivier Humbrecht
Domaine Zind-Humbrecht

4 ROUTE DE COLMAR, 68230 TURCKHEIM, FRANCE
WWW.ZINDHUMBRECHT.FR/EN

Olivier Humbrecht walks into any room and takes charge of it. It helps that he is tall and imposing physically of course, but it is more his confidence and razor-sharp intelligence that shines through. He might be tireless but he puts it to good use – not least by becoming the first ever French Master of Wine back in 1989. And he is always quick to defer to the brilliant people around him; I spent a fascinating few hours with his winemaker Paul McKirdy in Alsace last year, and came away having not only learned a ton of new things about Alsace terroir but so energized by his knowledge.

'We learn from others,' is how Olivier puts it. 'Working across generations is very positive – I worked with my father, and now with my son.'

Currently president of Biodyvin in France, he is a passionate advocate for other biodynamic growers and for the benefits of the system itself. It was the discovery that his soils gained ten times more caterpillars when he used biodynamics that convinced him to convert the entire vineyard in 1988.

'You're building the immune system of the soil and the vines, so they can counteract difficult vintages more easily.'

'What kills life in soils?' he asked at a recent conference. 'Easy: herbicides, chemicals, abusive soil compaction from heavy repetitive machine use, lack of corrective ploughings, not growing plants in the soils (young growing roots help to aerate a soil). When there is no more oxygen in a soil, life disappears. Just look what happens when someone always walk on the same track in a garden: it creates a path with no growth.'

'What brings life into a soil? Easy: use of high-quality compost, no chemicals or herbicides, allowing oxygen to penetrate into the soil by ploughing it in winter, allowing plants to grow by tilling and using lighter, less-damaging machines, and helping the soil digest all the fresh organic matter it gets every year by using specific preparations made from cow dung and plants: the basics for biodynamic cultivation.'
Oliver Humbrecht

DOMAINE ZIND-HUMBRECHT BRAND GRAND CRU RIESLING SELECTION DES GRAINS NOBLES 2008, FRANCE * * *

I nearly chose one of Olivier Humbrecht's dry Rieslings from the Rangen de Thann vineyard (try them!), but this is a beauty of a wine. Rich, fragrant, layered, spiced, dripping with honey, with 172g/l residual sugar it is cut through with such searing acidity that it just lifts off. Complexity doesn't even come close. Lots of flavours here for food matching, but can promise great results with a fig and mascarpone tart.

FRANÇOIS CHIDAINE LES LYS MONTLOUIS LOIRE 2009, FRANCE

WWW.FRANCOIS-CHIDAINE.COM

The leading producer in Montlouis-sur-Loire (along with Jacky Blot), François Chidaine is known for his skilled handling of Chenin Blanc. He makes stunning dry versions, but Les Lys shows why Chenin gives wines the freedom to be sweet and succulent without ever straying into heavy. Grown on clay and silex soils, underpinned by the famous tuffeau limestone, this is rich with apricot, peaches and nectarines, all fleshy and creamily textured over the steely core of Chenin's acidity (emphasized because it doesn't undergo the secondary softening malolactic fermentation). My advice with all sweet wines is to serve them as cold as you can. The freshness will give you more flexibility with food matches – try a plate of lightly cooked scallops.

HUGEL GERWURTZTRAMINER GROSSI LAÜE 2010, FRANCE

WWW.HUGEL.COM

Today Hugel is organic across 50% of their vineyard, with the rest on the way, but the vines for the steeply sloped marl soils of Grossi Laüe vineyard were the first to be converted. The family (Jean Frédéric Hugel working with uncles Jean-Philippe and Marc) produce wonderful Riesling from this vineyard, but the Gewurz is something a little different. It's the antithesis of a caricatured grape: it has the candied ginger, jasmine and almond, but none of the cloyingly sweet rose water that can be so head-spinning. The grapes are picked at full ripeness rather than with noble rot (or less than 5%) and it has just 16.3g/l residual sugar. You could almost believe it is dry, so concentrated is the citrus and so bracing and uplifting the acidity. Sink into the silkiness and the endless finish. Packed full of personality, more than a match for spicy food but also for foie gras or blue cheese.

WEINGUT WITTMANN MORSTEIN RIESLING AUSLESE 2015, GERMANY

WWW.WEINGUTWITTMANN.DE

The fresh, fragrant sweet Rieslings of Germany are complex and satisfying while being low in alcohol: fermentation is stopped halfway through the process, leaving some sugar behind. Günter Wittmann is a pioneer of organic winemaking in Germany, introducing it to his family vineyard in 1990. Today his son Philipp is winemaker, and rightly considered among the very best in the country. This wine is thrillingly lovely. Rippling through with delicate fresh peach and nectarine juice, gorgeously light and elegant – a brilliant match for a simple but delicious dessert such as whipped meringues or fruit tarts.

Overleaf: Alsace's Domaine Zind-Humbrecht brings life back to its soils.

Marie-Thérèse Chappaz
Domaine de la Colombe

CHEMIN DE LIAUDISE 39, 1926 FULLY, SWITZERLAND
WWW.CHAPPAZ.CH

The Queen of Valais, Icon of Swiss Wine, First Lady of Wine... you name the title, Marie-Thérèse Chappaz has been awarded it over the years. And yet she just might be the most low-key and self-effacing winemaker you can meet.

Born in Valais in 1960, she was given her first row of vines to look after (Pinot Noir) by her father when she was 17 years old. First specializing in viticultural research, she took over her family property in 1987, slowly piecing back together various plots of vines that had been mainly farmed out to other producers. And like so many of the winemakers in this book, she can trace the origins of her awakenings to more sustainable farming to one encounter – in this case with Michel Chapoutier in the Rhône back in 1997.

'I had never really heard of biodynamics before my trip to the Rhone, but when I saw that it was possible to be entirely natural in winemaking and possible to use no chemicals, it was like a light switching on. I had never liked using them but didn't realize there were other ways of doing things.'

'As soon as I got back to Switzerland I called (pioneering biodynamic consultant) François Bouchet and asked him to come look at my vines. I was thinking about converting one hectare, but he told me that was not enough. I started off with two hectares but by 2003 was fully biodynamic.'

'Treating half the vineyard traditionally was just wrong. It felt like I was giving one child great food and the other child fast food, it was impossible to continue like that.'
Marie-Thérèse Chappaz

DOMAINE CHAPPAZ GRAIN NOBLE PETITE ARVINE 2014, SWITZERLAND * * *

You start to understand why the wines from Marie-Thérèse Chappaz inspire such devotion when you open this bottle. It may be one of the most luscious sweet wines I've ever tasted. So dense and rich and then so soaring on the finish. Rich-amber in colour, on the palate apricot, bitter-orange marmalade, almonds and velvety-soft exotic fruits abound, but without ever getting heavy. The Petite Arvine grapes are from Domaine des Claives in Fully. It gets drier the longer you hold it in your mouth, and the flavours just stretch out endlessly. Wow, what a wine. Just begs for a plate of cheese – try an aged Gruyère with its slightly salty character and enjoy the contrast.

BERA VITTORIO E FIGLI CANELLI DOCG MOSCATO D'ASTI 2015, ITALY

WWW.BERA.IT

Tucked up in the hills of Piedmont, this heavenly, delicate and utterly delicious wine is pretty much ready-made for finishing off a meal. I'm a big fan of bringing back a sparkler to finish off supper, and this is a great choice for any fruit-based tart or a deliciously sticky apricot-up-side-down cake. The Bera family first got hold of vines in the Artesana hills in 1758, and today Alessandra and Gianluigi Bera work 10 ha, (organic since 1964). This is top-quality Moscato from Muscat Blanc à Petits Grains, fermented from wild yeasts, with no additives in the winemaking process, leaving the glass full of fresh peach blossom and soft apricot fruit flavours, with that gentle Asti sparkle.

Above: The Riesling grape is amazingly versatile, making wonderful wines from bone dry straight through to luscious and sweet.

Marinella Camerani
Corte Sant'Alda

VIA CAPOVILLA, 28 - LOCALITÀ FIOI,
37030 MEZZANE DI SOTTO, ITALY
WWW.CORTESANTALDA.COM/EN

Marinella Camerani describes farming organically and biodynamically as her way of 'paying back a debt that I have with this land'. She works with a small team at this Valpolicella estate, and has been fully organic since the 80s (converting 'a bit blindly' in 1985). However, she slowly became dissatisfied with it as a way of farming, until she attended Nicolas Joly's first-ever Italian conference on biodynamics back in 2002. 'I got a fresh jolt of motivation,' is how she tells it.

But today it is not just the benefits to the wine that count. Her colleague Matteo Piccoli Tosadori (who describes himself as working in the vineyard, winery, cellar, as a tour guide, you name it) says that the most important benefit has been for the team.

'For us, we see every day that the way we work has a direct impact on the wine and our environment. Seeing flowers and herbs alive in the vineyard is wonderful, but it has also brought us closer together.'

Working biodynamically means it is essential to be a team, to relate closely to each other and be aware of what is needed. So it is not just good for our sales, but for our spirits.'
Marinella Camerani

CORTE SANT'ALDA
Azienda Agricola
RECIOTO
Recioto della Valpolicella
DENOMINAZIONE DI ORIGINE CONTROLLATA E GARANTITA
• 2011 •

CORTE SANT'ALDA RECIOTO DOCG RECIOTO DELLA VALPOLICELLA 2011, ITALY

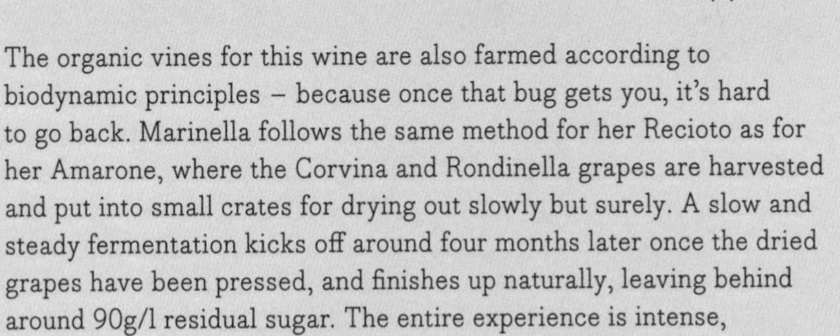

The organic vines for this wine are also farmed according to biodynamic principles – because once that bug gets you, it's hard to go back. Marinella follows the same method for her Recioto as for her Amarone, where the Corvina and Rondinella grapes are harvested and put into small crates for drying out slowly but surely. A slow and steady fermentation kicks off around four months later once the dried grapes have been pressed, and finishes up naturally, leaving behind around 90g/l residual sugar. The entire experience is intense, concentrated, complex flavours, and always a huge treat.

'Concentrated sweet wine of great harmony. Displaying delicious aromas of plum, liquorice, praline and some savoury touches. To partner a chocolate soufflé with a coconut ice cream.'
Gerard Basset MW, Best Sommelier in the World 2010

COSIMO MARIA MASINI FEDARDO DOC VINSANTO DEL CHIANTI 2009, ITALY ✱ ✱ ✱

WWW.COSIMOMARIAMASINI.IT

Another dream Italian location, the Cosimo Maria Masini winery is found on the hills around San Minato, known as the home of the Tuscan white truffle. The property has been owned by the Masini family since 2000, and combines olive groves and forests with vines. Besides this amazing wine, it is worth searching out the wines from indigenous varieties such as Buonamico and Sanforte. All wines are made with wild yeasts and left to progress at their own pace. The Vinsanto is made with 90% Trebbiano and 10% Malvasia Bianco, left to dry for four months then transferred into barrels for extremely slow fermentation and ageing that lasts five years. Its recommended pairing is with *cantuccini*, those beautiful tiny Italian almond biscuits. They are hard as a rock before you dip them into your Vinsanto, and then they melt away to nothing. A well-aged cheese will also do pretty well. With 109g/l residual sugar, you can honestly see why this is traditionally considered a holy wine; you can almost smell the incense curling out of the glass. You can easily lose yourself in the aromatics of this wine – by turns walnut, dried mandarin rind, figs, cardamom, saffron, hazlenuts, crème caramel, lime – so grab a chair by the fire and open a favourite book. An amazing wine.

FASOLI GINO ALTEO DOC AMARONE DELLA VALPOLICELLA 2011, ITALY

WWW.FASOLIGINO.COM/EN

Amarone is an experience like pretty much no other in wine. Prepare yourself if it's the first time you've drunk it, because in my experience the immediate reaction is to be uncertain, or even hostile. Its taste is so distinct because the grapes are late picked when fully ripe, and then left all winter to dry into raisins, which makes it both extremely concentrated and sweet, but with a bitter or astringent edge that comes from the drying process (Amarone comes from the word *amaro*, which means bitter in Italian). You can taste all of this in the glass – it is full of dried figs, scorched leather, tar, baked spices and balsamic, with a rich kick of liquorice on the finish. The Alteo, which comes from Fasoli Gino's best grapes, is made with a blend of Corvina, Corinone and Rondinella grapes, left in the drying loft for four months before fermentation, then aged in barrels for up to three years and in bottle for one. Achieving all of this takes time, patience and skill. My suggestion is to think of this like a coffee. Have it at the end of the meal with biscuits or chocolate and sink into the amazing balance of its sweet charm and powerful kick.

Herbert Zillinger
Weingut Herbert Zillinger

HAUPTSTRASSE 17, 2251 EBENTHAL, AUSTRIA
WWW.ZILLINGERWEIN.AT

Herbert Zillinger took over the family winery at the age of 20, back in 1998, after studying viticulture and fruit production at a local vocational high school, so it's probably fair to say that he has always been single-minded about what he wanted to do.

Today he runs the property with his speech-therapist wife Carmen, their two young daughters Amelie and Rosemarie and their dog Rust. And Herbert seems to inspire passion in plenty of wine-lovers – his is one of the names that kept coming up when I asked sommeliers about their favourite organic or biodynamic wines and producers. This industry recognition is really something to be proud of, as the estate was only certified in 2015 and joined the Respkt-BIODYN community of biodynamic wineries from Austria, Germany, Italy and Hungary in 2016. Alongside Demeter and Biodyvin, Respekt is now becoming another serious name in biodynamic wine and its members' bottles are definitely worth getting to know.

Herbert describes himself as 'a man without compromise', who follows his gut in the vineyard and cellar: natural fermentations, long ageing, minimum additions at any point... He also says not everyone likes the results and that he is 'not everybody's darling' – which is pretty much exactly the uncompromising attitude that you need to be a great winemaker.

WEINGUT HERBERT ZILLINGER PROFUND TRAMINER 2015, AUSTRIA

This is a beautiful and slightly head-spinning wine from the Traminer grape (sounds so much cooler than Gerwurtztraminer, yes?). It's a dry wine but believe me it works brilliantly in this section – spices, lychees, mandarins, dried fruits, rose petals, a sting of citrus all collide in your glass, but are held together by a tension and freshness. From a winemaker who is getting increasing recognition for what he is doing with grapes from the Weinviertel, the largest wine-producing region in Austria. Match to a sweet, spicy Cajun sauce or a sticky walnut tart.

DOMAINE KIRÁLYUDVAR TOKAJ CUVÉE PATRICIA LATE HARVEST 2012, HUNGARY

WWW.KIRALYUDVAR.COM

This estate was bought by the Hwang family of Domaine Huet in 1997, so a good number of years before they arrived in the Loire, and the estate's entire 45 ha has been converted to biodynamics. I strongly recommend trying the dry whites from this property, made with the Furmint grape. This is richer, with around 130g/l residual sugar, but the beauty of Tokaj is that the wines are always packed full of bitter-orange-rind flavours and sharp acidity to balance out the sweetness. With marmalade, honey, saffron and orange blossom, think saffron for food too: with grilled orange slices, or a flourless orange saffron cake.

PENDITS TOKAJI CUVÉE FURMINT HÁRSLEVELU 2012, HUNGARY

WWW.PENDITS.DE

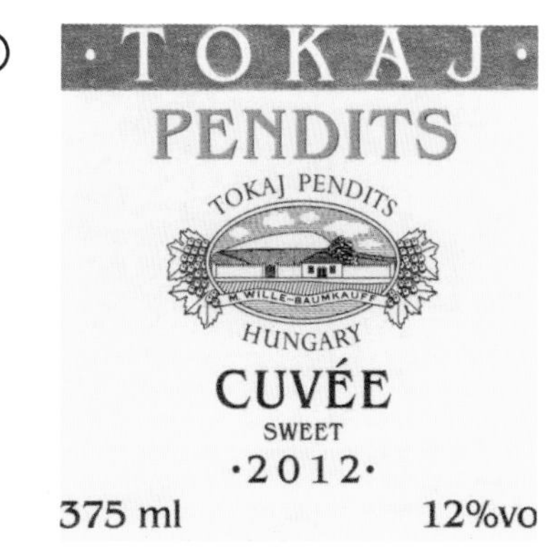

A great choice if you're not sure about sweet wine, because this is beautifully fresh and delivers more peach blossom and white peaches rather than overly sticky richness. It almost reminds me having of a Bellini in Venice, which has got to be a good thing. Owned by Márta Wille-Baumkauff and her son Stefan, Pendits is the name of an exceptionally high-quality vineyard in Tokaj where Márta owns 7 ha of vines. She is 100% biodynamic and 100% self-taught. This has 50g/l residual sugar and is made from an equal blend of Harslevelu and Yellow Muscat grapes. It would be brilliant served simply alongside a cup of coffee and a choux bun filled with crème pâtissière. Or go even more indulgent with a crêpe Suzette.

KALLESKE JMK SHIRAZ VP BAROSSA VALLEY 2015, AUSTRALIA

WWW.KALLESKE.COM

Hand-pruned and harvested by John Malcome Kalleske (hence the JMK), fifth-generation winemaker who has tended his family vineyards for over 50 years (I'm sure I'm not the only one to hope he had some help with the picking). This is a luscious, thoroughly indulgent wine. Packed full of mulberry and blackberry fruit, all wrapped in chocolate and cappuccino notes, with some tight tannins holding it all together and offsetting the sweet hit of the early palate. This was fermented in open vats with grape spirit added (as with Port) to stop the fermentation while there is still sweetness in the must. A brilliant match for a chocolate dessert, or salty blue cheese.

Ray Nadeson and Maree Collis
Lethbridge Wines

74 BURROWS ROAD, LETHBRIDGE, VIC 3332, AUSTRALIA
WWW.LETHBRIDGEWINES.COM

Any winemaker who states on the label that making this wine is a 'true financial folly' warms my heart. I live in a region where sweet wines are made from botrytis, or noble rot, infected grapes, and I know what they mean. I like to say masochists make this stuff and hedonists drink it.

But it's abundantly clear that Ray and Maree at Lethbridge are just those kind of people – honest, self-deprecating and totally committed. Both came to wine through a science background: Maree as a researcher in medicinal chemistry and Ray as a researcher and teacher of neuroscience at Monash University. For a while they combined both careers while launching Lethbridge with their friend Adrian Thomas who just happens to be a cardiovascular doctor (with Ray somehow also finding time to get a winemaking degree). Today they are full time here.

In fact, now I think about it these guys might just be the most highly qualified winemakers in the book. So it's probably no surprise that they apply the same vigour to sustainability – even the winery is built from bales of hay that provide a natural insulation and recreate, they say, 'the natural environment of underground cellars'. And I'd just like to confirm that I checked and rechecked that fact. 'Is your winery really built from bales of hay?' 'Absolutely,' said Ray. 'Designed and built with my own hands.' Even more amazingly, he did it back in 1999 and it's still standing.

2011 Lethbridge Botrytis Riesling
375mL

LETHBRIDGE WINES BOTRYTIS RIESLING 2011, AUSTRALIA

The three partners behind Lethbridge Wines all have medical backgrounds, but science takes a back seat to nature right from the moment you arrive here: olive trees at the gates, a sustainable, eco-friendly winery and cover crops and grasses growing between the vines. This was an extremely cool and wet vintage, but the pressure of rot meant that they could produce this beautiful Botrytis Riesling that combines gently teased-out exotic-fruit flavours with a stunning acidity that makes it almost dry on the finish. The fruit was sourced from a small two acre site in Henty in Western Victoria (a cool-climate region increasingly attracting plaudits for its sparkling and still white wines). Whole-bunch botrytis-affected grapes are pressed and wild fermented in barrels for 18 months until fermentation stops naturally. This is a beautiful wine, with a hauntingly soft lemon and papaya expression, and a gentle sweet-sour jolt that makes it a brilliant food wine – try something unusual like grilled pineapple and mint.

Above: Lethbridge Wines has farmed sustainably since its inception.

Glossary

SMALL CAPS cross-reference to other entries included in the Glossary.

ABV: Alcohol by volume, expressed as a percentage.

ALCOHOLIC FERMENTATION: The process of turning the sugar in wine grapes into alcohol, by the YEAST on the grape skins; the byproduct is CO_2.

AC/AOP: French classification term (Appellation d'Origine Contrôlée/Protegée). A certification granted to specific geographical indications for French wines and other agricultural products, with rules governing their production. These rules will still apply to wines made under ORGANIC or BIODYNAMIC FARMING in the relevant areas.

IGT/IGP: EU wine classification system (Indicazione di Geografica Tipica/Indication Géographique Protégée), which replaces the former Vin de Pays. Indicates that a wine comes from a specific area, but has less rules governing grape variety and production methods than AC.

BATONNAGE: A method used during wine production or ageing in either tank or barrel where the yeast lees are moved into suspension with the aim of increasing the body and mouthfeel of the resulting wine (see LEES STIRRING).

BIODYNAMIC FARMING: A practice of viticulture that rejects chemical and synthetic treatments, and works to bring balance to the ecosystem of a vineyard, aligning vineyard and cellar work with lunar and planetary movements. (See p.20.)

BLANC DE BLANCS: Champagne or sparkling wine made exclusively from white grapes, usually 100% Chardonnay.

BLANC DE NOIRS: Champagne made exclusively from red grapes, which in practice means either 100% Pinot Noir, 100% Pinot Meurnier or a blend of the two.

BODEGA: A wine cellar or wine-producing estate; a term most typically used in Spain or South America.

BOTRYTIS CINEREA: A fungus that affects wine grapes. In sweet wine production, known as noble rot, it differs from grey rot in that it doesn't destroy the grape, but allows it to become partially raisined. Makes concentrated sweet wine typically seen in Sauternes (France) and Tokaj (Hungary).

BRUT NATURE/BRUT ZERO: A Champagne or sparkling wine bottled without any DOSAGE or added sugar.

BRUT: A term used in sparkling wines and Champagne (dry), so with low levels of additional sweetness. Extra Brut and Ultra Brut are also seen, indicating even lower levels.

CARBONIC MACERATION: A specific form of WHOLE-BUNCH FERMENTATION used frequently in Beaujolais; grapes are left whole in bunches in a sealed vat that has been filled with CO_2, so ensuring no oxygen. The fermentation begins within the berries themselves before the skins split, encouraging extremely fruity flavours and low tannins.

CAVA: MÉTHODE TRADITIONELLE Spanish sparkling wine.

CHAPTALIZATION: The addition of cane or beet sugar to the wine before or during ALCOHOLIC FERMENTATION to increase the potential alcohol content. The addition of organic sugar or organic concentrated grape must is allowed in some circumstances in organic and biodynamic winemaking, although it is discouraged. It is not allowed, in theory, natural wines.

CHARMAT: A method of sparkling wine production where the wine undergoes a second fermentation in a pressurized tank, rather than in bottle. Most typically used in Prosecco.

CHÂTEAU-BOTTLED: On a French wine label meaning the wine has been produced and bottled on the same site.

CHEF DE CAVE: The cellarmaster in a Champagne house, responsible for blending.

CLOS: Historical term for a vineyard surrounded by walls.

CRÉMANT: MÉTHODE TRADITIONELLE French sparkling wine, made outside of the Champagne region.

CUVÉE: Typically refers to a specific bottling of a wine, separate to the main estate bottling.

DISGORGEMENT: Process of removing yeast sediments from a bottle of sparkling wine made using MÉTHODE TRADITIONELLE after the secondary fermentation in bottle.

DO/DOC(G): A controlled-quality wine region in Spain (Denominacion de Origen) and Italy (Denominazione di Origine Controllata/e Garantita). Portugal also uses DOCG.

DOSAGE: Sugar solution added to Champagne or sparkling wine after disgorgement to determine style.

DYNAMIZING: Or 'dynamic stirring', a method of introducing biodynamic preparations into water before their application in the vineyard. The technique creates a vortex through stirring the water first in one direction, then in the reverse, for around one hour. It can be done manually or by machine.

FATTORIA: A wine cellar or wine-producing estate, a term most typically used in Italy.

FILTRATION: A winemaking process where wine is passed through a filter to remove any particles from the liquid before bottling. The aim is to produce a clear and stable wine. Organic and biodynamic wines have specific rules over which products are allowed and many avoid it altogether. NATURAL WINES in theory allow no filtration (UNFILTERED).

FINING AGENTS: A product added at the end of the winemaking process to clarify or stabilize a wine through the removal of proteins and other organic compounds. Organic and biodynamic wines have specific rules over which products are allowed and many avoid it altogether. NATURAL WINES in theory allow no fining (UNFINED).

GG: Grosses Gewächs, or Great Growth in German wines.

GRAND CRU: Used in a variety of French wine classifications to indicate a wine or grapes of a superior quality. Most commonly in Champagne, where 17 villages have *grand cru* status. Only exception to this is in St-Émilion, where the term St-Émilion Grand Cru is simply an AC.

GROWER CHAMPAGNE: Sparkling wines made in the Champagne region where the grape-grower also bottles their own wine. Most big Champagne houses in contrast buy grapes from across the region.

LEES: The remains of yeast cells after ALCOHOLIC FERMENTATION. In theory, leaving the wine in contact with its lees adds complexity and body.

LEES STIRRING: Moving the yeast lees into suspension during the ageing of the wine (see BATONNAGE).

LIQUEUR DE TIRAGE: Sugar solution added at bottling to induce second fermentation in Champagne or sparkling wines.

LUTTE RAISONÉE: A middle path between chemical farming and strict ORGANIC FARMING; the intention is to apply as few vineyard treatments as possible (sustainable farming).

MALOLACTIC FERMENTATION: A secondary fermentation during winemaking, where the sharp malic acid found in the grapes is converted to a softer lactic acid through the action of naturally occurring or added bacteria. Lessens overall impression of acidity in the finished wine.

MÉTHODE TRADITIONELLE: A method of Champagne and sparkling wine production, inducing a secondary fermentation in bottle where the CO_2 is trapped in the liquid, producing bubbles.

METODO CLASSICO: Term used in Italy for sparkling wines made in the MÉTHODE TRADITIONELLE.

MONOPOLE: A vineyard (or vineyard area) owned by a single winery or wine company, most typically highlighted in Burgundy, where it is usual for vineyards to be split between several owners.

NATURAL WINE: Although there is no universally accepted definition or accreditation, typically natural wine is made with no additions in the process of winemaking, including no added yeasts, bacteria, enzymes, tannins and extremely minimal SO_2. (See p.26.)

ORANGE WINE: A white wine made by leaving grape skins and sometimes pips in contact with the juice, essentially making a white wine using a red wine technique. The resulting wine is darker coloured and more structured than a typical white wine. (See p.30.)

ORGANIC FARMING: Viticulture practiced with strict control over any treatments used in the vineyards, avoiding synthetic treatments, chemical pesticides, herbicides and fertilizers. (See p.16.)

PÉT-NAT: Pétillant Naturel, a naturally sparkling wine, made by bottling still-fermenting wine and capping it, naturally trapping the CO_2, which is a byproduct of the fermentation process, which creates bubbles.

PH: A chemical measurement of the intensity of the acidity in a wine; the lower the pH, the more intense the perception of acidity.

PUMPING OVER/PUNCHING DOWN: The colour in wine comes from the skins being in contact with the juice during winemaking. These techniques ensure this happens, so either by punching the skins down into the juice, or by pumping the juice over the skins.

QUINTA: A wine cellar or wine-producing estate, a term most typically used in Portugal.

QUALITÄTSWEIN MIT PRÄDIKAT (QMP): A German classification of wine, meaning 'quality wine with specific attributes' (or Prädikatswein). Wines are ranked according to the ripeness level of the grape when harvested. From lowest to highest: Kabinett, Spatlese, Auslese, Beernauslese (BA) and Trockenbeerenauslese (TBA).

REDUCTION/REDUCTIVE/REDUCED: All slightly different terms talking about the amount of oxygen in the wine or winemaking process. Reductive winemaking involves limiting the contact with oxygen to put the emphasis on fresh fruit flavours. If there is too little oxygen, a wine can become reduced, which is a fault, and will display burned rubber or struck match flavours... or rotten eggs if really extreme. An area with lots of ongoing debate.

RESIDUAL SUGAR: A measurement, usually expressed in grams per litre (g/l), of the amount of grape sugar remaining in a wine after ALCOHOLIC FERMENTATION is completed. Dry wines have little or none; dessert wines have varying levels.

ROOTSTOCK: The roots on which a vine plant grows. Today, most *Vitis vinifera* vines are not planted on their own roots but are grafted on to the root of a different vine species.

SKIN-CONTACT: Leaving the grapes in contact with their skins during winemaking.

SOLERA: The system of fractional blending multiple vintages of a wine during the ageing process.

SULPHUR DIOXIDE (SO_2): The main tool for wine stabilization as both an antioxidant and an antiseptic/anti-microbial agent. Organic winemakers aim to keep their use of SO_2 to a minimum, and natural winemakers often dispense with it altogether. Allowable levels differ depending on country and certification body.

SUPER TUSCAN: 'Unofficial' (typically outside regulations) red wine from Tuscany made using international grape varieties, most usually Cabernet Sauvignon and Merlot.

TERROIR: A French term (in origin at least) that refers to the specific soil, topography, climate, landscape character and biodiversity features, including the influence of man, of a named place where wine is grown. It is believed to influence taste and character of a wine, and is an important concept in both ORGANIC and BIODYNAMIC FARMING.

UNFILTERED: A wine that has been bottled without going through the process of filtration. See FILTRATION.

UNFINED: A wine that has been bottled without going through the process of fining. See FINING AGENTS.

UNGRAFTED VINE: Vines grown on their original ROOTSTOCK. Also known in French as *franc de pied*.

VEGAN WINES: Animal byproducts may be used as FINING AGENTS to clarify wine, including egg white, casein (a milk protein), gelatine and isinglass (derived from fish). A vegan wine uses no animal products whatsoever.

WHOLE-BUNCH FERMENTATION: A winemaking technique where grapes are left intact, still attached to their stems (also known as whole cluster), before being put into a vat for fermentation. In contrast to the more typical process where stems are removed entirely before winemaking.

YEAST: Micro-organism found naturally on grape skins, causes ALCOHOLIC FERMENTATION. Specially selected yeasts may be used, but this is unusual in organics, rare in biodynamics and, in theory, forbidden for NATURAL WINES.

Directory

WHERE TO BUY THE WINES

AUSTRALIA

Dan Murphy's: www.danmurphys.com.au
Dynamic Wines: www.dynamicwines.com.au
Organic Wholefoods: www.wholefoods.com.au
Organic Wine Store: www.organicwine.com.au
The Organic Cellar: www.theorganiccellar.com
The Organic Wine Cellar:
www.theorganicwinecellar.com.au
Veno: www.veno.com.au

FRANCE

Au Bon Vingt:
www.facebook.com/AU-BON-VINGT-1627078234220175
Cave Legrand Filles et Fils:
www.caves-legrand.com
Gevrey Wine Club: www.gevreywineclub.com
Lavinia: www.lavinia.fr
Ma Cave Fleury:
www.macavefleury.wordpress.com
Petits Caves: www.petitescaves.com

HONG KONG

121BC: www.121bc.com.hk
La Cabane: www.lacabane.hk
Simply Wine: www.simplywines.com.hk

IRELAND

Curious Wines: www.curiouswines.ie
Le Caveau: www.lecaveau.ie

NEW ZEALAND

Accent on Wine: www.accentonwine.co.nz
Organic Wine Company:
www.organicwinecompany.co.nz
Sapori d'Italia: www.sapori.co.nz

SOUTH AFRICA

Vino Pronto: www.vinopronto.co.za
Wine Cellar: www.winecellar.co.za

THE UK

Armit Wines www.armitwines.co.uk
Aubert & Mascoli: www.aubertandmascoli.com
Berry Bros. & Rudd: www.bbr.com
Bottle Apostle: www.bottleapostle.com
Buon Vino: www.buonvino.co.uk
Hanging Ditch: www.hangingditch.com
Indigo Wines: www.indigowine.com
Les Caves de Pyrene: www.lescaves.co.uk
Meadowdale Ecological Wines: www.meadowdalewines.com
Private Cellar: www.privatecellar.co.uk
The Wine Society: www.thewinesociety.com
Uncorked: www.uncorked.co.uk
Vinceremos Organic Wines: www.vinceremos.co.uk
Vintage Roots: www.vintageroots.co.uk

THE US

Brooklyn Wine Exchange:
www.brooklynwineexchange.com
Candid Wines: www.candidwines.com
Chambers St: www.chambersstwines.com
Domaine LA: www.domainela.com
Flatiron: www.flatiron-wines.com
Frankly: www.franklywines.com/password
Leon&Son: www.leonandsonwine.com
Terroir: www.terroirsf.com
The Natural Wine Company: www.naturalwine.com
Wholefood Markets: www.wholefoodsmarket.com

WHERE TO EAT & DRINK

AMSTERDAM

Bar Centraal:
www.facebook.com/barcentraalamsterdam
Café de Klepel: www.cafedeklepel.nl/?lang=en

AUCKLAND

The Grove: www.thegroverestaurant.co.nz

BORDEAUX

Le Pointe Rouge: www.pointrouge-bdx.com

CAPE TOWN

Chalk & Cork: www.chalkandcork.co.za
Chef's Warehouse: www.chefswarehouse.co.za
Publik Wine: www.facebook.com/publikwine

COPENHAGEN

Amass restaurant: www.amassrestaurant.com
Den Vandrette wine bar: www.denvandrette.dk
Manfred & Vin: www.manfreds.dk
Relae restaurant: www.restaurant-relae.dk/en
Terroiristen wine bar: www.terroiristen.dk

DUBLIN

Ely Wine Bar: www.elywinebar.ie

EDINBURGH

Good Brothers: www.goodbrothers.co.uk

HONG KONG

Le Quinze Vins : www.facebook.com/LQVHK
My House: www.facebook.com/MyHouseHongKong
Serge et le Phoque: 5465 2000

KRISTIANSAND (NORWAY)

Bolgen & Moi: www.bolgenogmoi.no

LONDON

67 Pall Mall: www.67pallmall.co.uk
40 Maltby Street: www.40maltbystreet.com
Chiltern Firehouse: www.chilternfirehouse.com
Noble Rot Wine Bar: www.noblerot.co.uk/wine-bar
Primeur: www.primeurn5.co.uk
Sager + Wilde: www.sagerandwilde.com
Terroirs Wine Bar: www.terroirswinebar.com
The Remedy: www.theremedylondon.com
The Winemakers Club : www.thewinemakersclub.co.uk

LOS ANGELES

AOC: www.aocwinebar.com
Bar Bandini: www.barbandini.com
Mignon: www.mignonla.com

MELBOURNE

Embla: www.embla.com.au
Gerald's Bar: www.geraldsbar.com.au
Host: www.hostdining.com.au

NEW YORK

Aldo Sohm wine bar: www.aldosohmwinebar.com
Compagnie des Vins Surnaturels:
www.compagnienyc.com
DOC Wine Bar: www.docwinebar.com
Rouge Tomate: www.rougetomatechelsea.com
The Ten Bells: www.tenbellsnyc.com
The Waverly Inn : www.waverlynyc.com
Wildair : wildair.nyc

PARIS

Aux Deux Amis:
www.facebook.com/Aux-Deux-Amis-120307971356817
Chez les Anges: www.chezlesanges.com
Fines Gueules: www.lesfinesgueules.fr
La Buvette:
www.facebook.com/La-Buvette-141341299357225
La Crémerie: 01 43 54 99 30
Le Baratin : 01 43 49 39 70
Racines Paris: www.racinesparis.com
Septime La Cave: www.septime-charonne.fr
Vivant Paris: www.vivantparis.com

PROVENCE

Château La Coste: www.chateau-la-coste.com

SAN FRANCISCO & BAY AREA

Culinary Institute of America, Napa:
www.ciachef.edu
Green Ellen Star: www.glenellenstar.com
Press Restaurant: www.pressnapavalley.com
Yield Wine Bar: www.yieldandpause.com

STOCKHOLM

Grand Hotel Stockholm Wine Cellar:
www.grandhotel.se/en
Mathias Dahlgren: www.mdghs.se/en

SYDNEY

10 William Street: www.10williamst.com.au
Bloodwood: www.bloodwoodnewtown.com
Est: www.merivale.com.au/est
Monopole: www.monopolesydney.com.au

Picture Credits

Front Cover: Kim Lightbody (© Jacqui Small)
Endpapers: Courtesy of Nikolaihof of Wachau & Podere Forte vineyards
Page 1: Courtesy of Lightfoot & Wolfville vineyards
Page 3: Janis Miglavs, courtesy of Cain Vineyard & Winery
Page 4: Theo Zierock
Page 7: Courtesy of Lethbridge Wines
Page 8: Eirik Sand Johansen for G.D. Vajra
Page 9: Raventós i Blanc
Page 10: Kim Lightbody (© Jacqui Small)
Page 13: Kim Lightbody (© Jacqui Small)
Page 14-15: Kim Lightbody (© Jacqui Small)
Page 17: Kim Lightbody (© Jacqui Small)
Page 18: Harry Annoni
Page 19: Harry Annoni
Page 21: Kim Lightbody (© Jacqui Small)
Page 22: Klagyivik Viktor/Shutterstock.com
Page 23: INTERFOTO/Alamy Stock Photo
Page 24: Raventós i Blanc
Page 25: F. Nivelle
Page 27: Kim Lightbody (© Jacqui Small)
Page 28: Imanol Legross
Page 29: Raventós i Blanc
Page 31: Kim Lightbody (© Jacqui Small)
Page 32: Courtesy of Meinklang vineyard
Page 33: Peter Podpera, courtesy of Weingut Hirsch winery
Page 35: Harry Annoni
Page 36: Courtesy of Château La Fleur
Page 38: Eric Zeziola
Page 39: Theo Zierock
Page 41: John Carey
Pages 42-43: Kim Lightbody (© Jacqui Small)
Page 44: Anne-Emmanuelle Thion
Page 47: Courtesy of Gramona wjacinery
Page 48: Raventós i Blanc
Page 49: Marçal Font
Pages 50-51: Raventós i Blanc
Page 53: Courtesy of Champagne Fleury
Pages 54 & 55: Courtesy of Champagne Marguet
Pages 58 & 59: Eric Zeziola
Page 60: Courtesy of Jacques Selosse
Page 63: Courtesy of Matetic vineyard
Page 64: Courtesy of Château de Béru
Page 66: Courtesy of Domaine de la Boissonneuse
Page 68: Courtesy of Weingut Peter Jakob Kühn
Page 69: Andreas Durst
Page 71: Courtesy of Nikolaihof of Wachau
Pages 72 & 73: Peter Podpera, courtesy of Weingut Hirsch winery
Page 75: Courtesy of Domaine Spiropoulos
Pages 78, 79 & 80-81: Kim Lightbody (© Jacqui Small)
Page 82: Courtesy of Franck Moreau MS
Pages 84 & 85: Courtesy of Frog's Leap winery
Pages 86 & 87: Rocco Ceselin
Page 88: Courtesy of Vale Da Capucha vineyard
Page 89: Obrigado Vinhos
Page 91: Courtesy of Palacio de Canedo
Pages 92 & 93: Carlos Spottorno
Page 94: Courtesy of Dom. Alexandre Bain
Page 96-97: Courtesy of Château Romanin
Pages 98 & 99: Courtesy of Coulée de Serrant vineyard
Pages 100 & 101: Claude Cruells
Page 102: Guillame Atger
Page 103: Courtesy of Domaine Ganevat
Page 104-05: Rocco Ceselin
Pages 106 & 107: Courtesy of Dom. Leflaive
Page 108: Courtesy of Domaine Ostertag
Page 113: Courtesy of La Stoppa vineyard
Pages 114-15, 116 & 117: Fabio Rinaldi
Page 119: Courtesy of Massena vineyard
Page 121: Courtesy of The Islander winery
Page 122-23: Kim Lightbody (© Jacqui Small)
Page 124: Wendy Wilmoth
Page 125: Courtesy of Lightfoot & Wolfville vineyards
Page 126: Courtesy of Corison winery
Page 128-29: Courtesy of Lightfoot and Wolfville vineyards
Page 130: Courtesy of Domaine Drouhin
Page 132: Sara Remington
Page 134: Covadonga Valdueza
Pages 136 & 137: Anaka Photography
Page 140: Courtesy of Domaine de la Romanée-Conti
Page 141: Courtesy of Domaine des Roches Neuves
Page 142: Courtesy of Domaine de Trévallon
Page 143: Harry Annoni
Page 144 & 145: Courtesy of Domaine Henri & Gilles Buisson
Page 148: Courtesy of Domaine des Chênes
Pages 150, 151 & 152-153: Courtesy of Domaine Stéphane Ogier
Pages 154 & 155: Courtesy of Weingut Ökonomierat Rebholz
Pages 155 & 156: Theo Zierock
Pages 158 & 159: Eirik Sand Johansen for G.D. Vajra
Page 160: Courtesy of Occhipinti winery
Page 163: Courtesy of Montevertine winery
Page 164: Courtesy of Intellgo winery
Pages 166 & 167: Courtesy of Wang Zhong winery (Tinsai vineyards)
Page 168: Courtesy of Ochota Barrels winery
Page 169 & 170-171: Courtesy of Seresin vineyards
Page 174-75: Kim Lightbody (© Jacqui Small)
Page 176: Courtesy of Ronan Sayburn MS
Pages 178 & 179: Charles O'Rear, courtesy of Cain Vineyard & Winery
Page 180: Courtesy of Cowhorn vineyard
Page 182-83: Courtesy of Benziger Family Winery
Page 184: Courtesy of Cooper Mountain Vineyards
Page 186: Courtesy of Maysara Wines
Page 189: Courtesy of Bodega Colomé
Page 192: Courtesy of Dominio de Pingus
Page 193: Carlos Gonzalez Armesto
Page 194: Courtesy of Mas La Plana
Page 195: Jordi Elias
Page 196: Courtesy of Tentenublo Wines
Pages 198 & 199: Courtesy of Château Lafleur
Page 200: Courtesy of Château de Puy
Page 201: Antoine De Tapol
Page 202: Courtesy of Château Pontet-Canet
Pages 203 & 204-05: Laurent Pamato
Page 206: Courtesy of Domaine de Fondrèche
Page 208: Courtesy of Clos d'Ora/Gérard Bertrand
Page 210: Courtesy of Mas de Daumas Gassac
Page 212: Courtesy of Fontodi
Page 214: Courtesy of Pian dell'Orino
Pages 216 & 217: Courtesy of Sesti winery
Page 220: Courtesy of Ao Yun winery
Page 221: Courtesy of Bekkers Wine
Page 223: Courtesy of Yangarra Estate
Page 224-25: Kim Lightbody (© Jacqui Small)
Page 226: Courtesy of Robert Bath MS
Page 229: Courtesy of The Fladgate Partnership (Fonseca)
Page 230: Vincent Bengold
Page 231: François Poincet
Page 232: Courtesy of Château Guiraud
Page 235: Courtesy of Domaine Huet
Page 236: Courtesy of Dom. Zind-Humbrecht
Page 238-239: Norbert Hecht
Page 240: Dominique Derisbourg
Page 241: Bon Appetit/Alamy Stock Photo
Page 242: Courtesy of Corte Sant'Alda
Page 244: Courtesy of Weingut Herbert Zillinger
Pages 246 & 247: Courtesy of Lethbridge Wines

Index

Bibliography & Further Reading

BIBLIOGRAPHY

Aron, Jules, *Zen and Tonic* (New York, Countryman Press, 2016)

Bertrand, Gérard, *Wine, Moon and Stars: A South of France Experience* (New York, Abrams Books, 2015)

Fukuoka, Masanobu, *The One Straw Revolution, An Introduction to Natural Farming* (New York, New York Review Books Classics, 2009)

Heekin, Deirdre, *An Unlikely Vineyard* (Vermont, Chelsea Green Publishing, 2009)

Howard, Sir Albert, *An Agricultural Testament* (Oxford, Benediction Classics, 2010)

Joly, Nicolas, and Benzinger, Mike, and Greene, Joshua, *Biodynamic Wine Demystified* (San Francisco, Board and Bench Publishing, 2008)

Steiner, Rudolf, *Agricultural Course: The Birth of the Biodynamic Method* (East Sussex, Rudolf Steiner Press, 2004)

Waldin, Monty, *Biodynamic Wine* (London, Infinite Ideas, 2016)

FURTHER READING

Association des Vins Naturels (AVN): www.lesvinsnaturels.org

Biodyvin: www.biodyvin.com

Bio-Siegel: www.oekolandbau.de

CCOF Organics: www.ccof.org

Certifé Agriculture Biologique: www.agencebio.org

Demeter: www.demeter.fr; www.demeter-usa.org; www.demetercanada.ca

Ecocert: www.ecocert.com/en

Claude and Lydia Bourguignon: www.lams-21.com

NASAA Organic: www.nasaa.com.au

Organics International: www.ifoam.bio

Organic labels around the world: www.organic-bio.com/en/labels

Pierre Masson: www.biodynamie-services.fr

Respekt Biodyn: www.respekt-biodyn.bio/en/

Rodale Institute: www.rodaleinstitute.org

Soil Association: www.soilassociation.org

USDA Organic: www.usda.gov

Acknowledgements

This book was truly a joint effort. I could not have completed it without the invaluable help and advice from many brilliant sommeliers, consultants and colleagues who shared their choices of favourite and most-admired wine producers. These include Franck Moreau, Bob Bath, Pascaline Lepeltier, Gerard Basset, Ronan Sayburn, Jeff Harding, Philippe Jamesse, Jules Aron, Caleb Ganzer, Simon Woolf, Toby Bekkers, Claude and Lydia Bourguignon, Andreas Wickhoff, the team at the Wine Society, Berry Bros. & Rudd, Indigo Wines, Noble Rot, h2Vin and many others for their inspiration and advice.

A special thank you to Doug Wregg at Les Caves de Pyrene for organizing tastings and lending us your beautiful Terroirs wine bar for the photo shoot. Also thanks to the many producers who gave me their time in explaining the philosophy behind their farming, and who shared their wines with me – with a special sorry to Podere Forte for sending their amazing Sangiovese that we finally just didn't have the time to include – wonderful wine, thank you. Tasting help came from Caroline and Charlie Matthews, Jo Simpson, Dan Snook, Shaun and Julie Bishop, Nikki and Jules Garafano, Sophie Kevany, Lorraine Carrigan, Lars Venborg and many other friends along the way. I have to also thank Monty Waldin for his fantastic books on biodynamics that should be widely read by anyone interested in further exploring the subject.

A big thanks to Amanda Skinner for putting me in touch with Jacqui Small for the early discussions on the book – and to the brilliant and tireless Fritha Saunders, Joe Hallsworth, Emma Heyworth-Dunn, Luke Fenech and Hilary Lumsden for all the production, design and editorial work. Thanks to Elin McCoy, Victoria Moore, Michel Roux and Francis Mallmann for reading early proofs and giving feedback. And finally to Francis, Lauren and Emily, as always.